Water on Earth

The Story of its Origin, Habitats, Neglect and Regeneration

SHRUTI MATHUR
RAJENDRA KUMAR

RAWAT PUBLICATIONS
Jaipur • New Delhi • Bangalore • Guwahati • Kolkata

ISBN 978-81-316-0900-2

Published by
Prem Rawat for **Rawat Publications**
Satyam Apts, Sector 3, Jawahar Nagar, Jaipur 302 004 (India)
Phone: 0141 265 1748 / 7006 Fax: 0141 265 1748
E-mail: info@rawatbooks.com
Website: www.rawatbooks.com

New Delhi Office
4858/24, Ansari Road, Daryaganj, New Delhi 110 002
Phone: 011 2326 3290

Also at *Bangalore, Guwahati* and *Kolkata*

Typeset by Rawat Computers, Jaipur
Printed at Chaman Enterprises, New Delhi

Dedicated to

Kusum
mother of Shruti Mathur and wife of Rajendra Kumar

It was through her that one of us had the privilege of witnessing the Divine Aura of
Maa Anandmayee

Contents

Foreword

Accepting water as a divine gift, and granting its transcendental flow a status of social reverence, on the one hand, and accepting the challenges arising out of its anticipated shortages in the 21st century; on the other Dr Shruti Mathur and Dr Rajendra Kumar, of Indian intelligentsia, with their deep understanding of the complex issues involved, made a well organized and systematic presentation in simple language of the complex scientific issues involved to expound the decentralized nature of the socio-scientific issues connected with the continued availability of clean water.

I regard the book as providing scientific directions to avert the gathering of dark and threatening clouds, which portent the third world war (for gaining control of water resources of the world), and also as one that creates new awakening to make water plentifully available to all sections of the society. I commend the authors for spreading water literacy amongst water management bureaucrats, social workers, politicians and general public alike.

I hope that the book succeeds in changing the mindset of people and creates moral awakening. It is heartening that the book, after making aware of the threatening issues, provides directions to meet and overcome the emerging vexed problems.

I congratulate both the authors for the comprehensive efforts to spread scientific awareness about water. Wish Dr Shruti Mathur and Dr Rajendra Kumar successful and healthy life.

Rajendra Singh
Jal-Purush (Water Man of India)
Winner of Nobel Prize for Water (2015)

Preface

This preface should have been written by Shruti as she is the prime author of the book, but respectfully she left that job to her father.

As an eager university scholar in Bhopal, Shruti, the main author of the book, was greatly concerned and agitated over many societal disparities; one amongst them was dwindling resources of fresh water and the consequent scarcity of water. Whilst the authors had reasonable supply of water in their government bungalow, others were not so blessed, particularly those living in *jhuggies* or the shanties in the so-called slum areas. Women there had to queue up for hours together before a single municipal tap which would become live only for a limited period of time, hardly sufficient for each family to collect about 20 litres of water to meet their domestic requirements. She was both concerned and disturbed on seeing fresh water being liberally used to water the lawns in the affluent houses or cars/bikes being washed with torrents of water streams from bore wells. She had earlier heard from me how the well in our home in Lucknow had progressively dried up due to receding water table.

By the time she moved to Jaipur to start her innings in the academic profession, water scarcity had become acute in many parts of India, including Rajasthan. Even as massive dams were being built all over the country to help irrigate the fields, farmers were committing suicides because of the failure of their crops due to scarcity of water. The science she knew always cautioned her that the scarcity was not due to any cataclysmic earthly event.

Shruti had not forgotten my close association with the scientific issues relating to environment and global warming. I was the only Indian scientist invited by the United Nations Centre for Science and

Technology for Development, New York to participate in its workshops at Moscow (1989) and Saarbrucken, Germany (1990–91) which were a prelude to the 1992 Rio Summit on global warming. A young Indian engineer (who could peer into future) had then foretold me that is should be working in the field of 'Water'. Incidentally, as other predictions made by him had proved correct, including his description of the venue of the Conference at Saarbrucken, Shruti began to pressurize me to join her in her efforts to create social awareness about water by educating the masses scientifically. For both of us, whose professional life was, and is, devoted to academic interests, there was no other way of doing so than through a comprehensive book, dealing with various facets of water, from its origin from saline oceans, the appearance of soil on the otherwise rocky planet, its natural habitats, its utilization patterns and global distribution and the effects of human activities on its habitats.

She had learnt in her school geography that the Arab peninsula was an extension of the harsh Sahara Desert. She would look in awe when her uncle, working as an engineer in United Arab Emirates, told her on his annual home visits that they have abundant supplies of water there and plenty of irrigated greenery through desalination sea water and scientific management of water resources. His glorification of water situation in a country which her school geography books had categorized as a water deficient desert, created in her youthful mind that the situation in India is not beyond redemption, if the common man is equipped with scientific knowledge, shorn of dogmas, through a book. I succumbed to her pressures and agreed to reinforce her efforts only on the condition that she would take the major load on herself, with me acting only as an advisor or moderator.

This is the background for the book. We begin the book by reminding people of our rich cultural traditions of reverence for water and go on to describe in simple language how nature created fresh water from the saline sea and had to craft habitats on earth to store fresh water. The pristine nature of these habitats was gradually getting destroyed by anthropogenic activities and societal indifference towards on account of lack of knowledge about them. All this had resulted in neglect of resources and abusive use of water. The net effect of all this is that the sword of Damocles is hanging over the global society that unless the situation is reversed soon, there would be grim shortage of water by the year 2030.

This book is primarily written to create awareness in the society, particularly amongst the youth so that the looming threats of destruction of human society are averted.

We feel deeply honoured as Shri Rajendra Singh, globally honoured and respected man, hailed as the 'Water Man of India', *or Jal Purush,* for the Foreword he has written for the book.

Since the book is meant for societal advancements, Shruti and I have decided that all royalties earned through the sale of the book would be humbly passed on to Sree Sree Maa Anandmayee Charitable Hospital, Varanasi, India.

Rajendra Kumar

Acknowledgements

A number of people have helped us in our task of writing the book. I/We can mention only a few, who are: Gopendra Kumar, Retired Director Geological Survey of India (Northern Circle), Lucknow; N.K. Nag, Engineering and Management Consultant, for critically reviewing the contents of the book and for many useful suggestions Ranjana Nag, noted teacher of English Language and Jhilmil Nag (Microsoft, Singapore) and Aishwarya Mathur, engineering student, PES University, Bangalore, for teaching Rajendra Kumar the use of computer for writing.

Rajendra Kumar would also like to thank Dr. Anurag Kumar for keeping him in good health to complete the task and hopefully beyond. Shruti Mathur would like to express her gratitude to the authorities of Amity University, Jaipur, for providing her an academic environment and encouragement for academic excursions.

The authors would also like to thank Shri Pranit Rawat for his wise counsels and advice.

Overriding all these, the primary source of inspiration for writing this book has been Smt Kusam Mathur whom Maa Durga recalled to be in her service in July 2016. No words can express our gratitude to her.

Shruti Mathur
Rajendra Kumar

Chapter 1

Introduction

Water is basic for the survival of humans and their civilization on planet earth. All of the habitats of water on earth were crafted by Nature over millions of million years ago, either by violent geological activities or gradually through breakdown of rocks through forming faults and cracks, erosion and weathering of rocks and fluvial action turning them into gravel, sand, sediments and alluvium. Important amongst the habitats are: underground aquifers in alluvial soil or in fragmented sedimentary rocks as well as those on the surface in rivers, springs, ponds and lakes, polar ice caps and glaciers including permafrost. These are all under acute stress as a result of human activities. In this regard, even the primal source of them all – the rains – is not spared as the nature of rainfall has changed in terms of its timing, distribution and quantity of precipitation.

Until about two-and-a-half centuries back, fresh water was plentifully available with almost guaranteed sustainability in its natural habitats; whatever quantity of water came out of them continued to be replenished through the annually repetitive hydrological (water) cycle, which has been running since aeons with remarkable precision. Its consumption during this historically short period, specially during the last 6–7 decades, has tremendously increased, on account of huge increase of human population, per se, and in irrigating the expanding farm field areas in order to provide mankind food security as also in variety of mineral-based chemical and engineering industries. This period simultaneously witnessed growth of electric power generation through the combustion of fossil fuels which also pushed up the demand for fresh water as well. The consumptions of energy and water came soon to be regarded as the two sides of the 'coin' called progress. Together, they have brought the world to the brink of disaster – the former through causing global warming and the latter through unmindful excessive or unsustainable consumption resulting in over exploitation which often caused irrevocable destruction of natural habitats or resources of water.

The per capita availability of water in different regions of the world shows wide disparities. In many of the tropical and heavily populated countries, its per capita availability is already less than the minimum considered necessary at 1,000 cu. m per annum by the World Health Organization for ensuring healthy life with adequate food security. Estimates of the demographic experts that the world population would exceed 9 billion by 2050 and that the increase would be most pronounced in the sub-tropical countries has unnerved global water management gurus.

While the situation in India as a unit is not so bad as in some other parts of the world, there is worrisome regional inter-regional disparity. Whilst some regions are blessed with its abundance, some others have reasonable supplies, yet there are regions facing acute shortages where its per capita availability is about a third of the globally accepted minimum. This fact alone points to the need of creating scientific awareness about water so that the present age Indians (as indeed the rest of the world too) restore the reverence which India has traditionally been giving to water and its resources. Centuries ago, India was the first country in the world to have introduced water management as an issue of state concern.

Geographically, the resources and habitats of water are inequitably distributed across the world – the most populous China and other countries of South-East Asia are less endowed but those with small populations, such as Bhutan in our immediate neighbourhood, are more than well endowed. The per capita availability of water in Pakistan and Bangladesh is not happy either. We share river waters with them. Since global warming affects through atmosphere which is not influenced by political boundaries, the world is united in bracing itself to mitigate global warming. Countries are, however, ill-prepared to share global water resources equitably. Water, unlike air, is unable to break barriers of national identities and self-interest. Many countries are unilaterally trying to grab waters of international rivers within their geographical confines, unmindful of the consequences of their unilateral actions on downstream countries. Global water resources may soon have to be regarded as a common resource for the entire human population if we are to avert wars. Water issues will necessarily have to be managed at country level with understanding that no country will dam the internationally flowing rivers for its exclusive benefit without regard to the needs of countries in its downward passage to the sea.

What contrasts the two looming threats to human existence on Earth (those of global warming and water scarcity) is the fact that whereas global warming would affect the society as a whole, that of water shortages would affect at individual and personal levels, even within a country. Experts visualize that the shortages would affect individuals much more cruelly than the industry. National and international experts, along with those of international organizations such as the UN, UNESCO, etc. foresee that the water shortages would assume critical dimensions in foreseeable future, i.e. 2025–2050. Further what makes the issue really gruesome is the fact that it would selectively target the economically weaker sections of the society. In India alone, this works out to be more than 700 million of human heads, and on global scale to be more than 3,500 million. Their very survival would be at stake through non-availability of drinking water and water to facilitate sanitation or to provide food security. Since tentacles of water scarcity would reach individuals, it could lead to internecine class wars. Inter-state feuds in India, some even needing judicial intervention at the level of the Supreme Court are not uncommon.

Scientists foresee that a bad situation turning worse with global climate change because it would profoundly affect the hydrological cycle through melting of ice caps, shrinking of glaciers and rising sea levels. Already, the growing tendency towards acidification of alkaline sea water has threatened the very existence of corals and coral reef islands. Unless the trend is reversed, it could adversely affect the habitats of different types of marine fish.

Incidentally, the major petroleum rich countries are water and rainfall deficient. Their continuing search for additional deposits of petroleum in the second half of the 20th century led to exploration reaching continually increasing depths in the bowels of the Earth which also unfolded the presence of aquifers with what then appeared to hold abundant stocks of water. According to the geologists, water in these confined aquifers had accumulated thousands of years ago and remained confined there, until now. These countries, prominently USA (California), Saudi Arabia, Libya besides others, started pumping up huge quantities of water in the last two decades of the 20th century.

As a result, California became fruit basket of USA, desert-bound Saudi Arabia started cultivation of wheat on large scale to become the sixth largest exporter of wheat in the world, till about a decade back. Libya exceeded all limits and created an engineering marvel in the form of a man-made river on the strength of underground aquifers, the geographical spread of which crossed international boundaries. Unilateral withdrawal of unrestricted quantities of water, without any recharge, has already emptied some of the aquifers within a short span of just a few decades. A Saudi report on the subject had aptly summed up the situation by stating: 'You cannot make camels fly or deserts bloom'. Saudi Arabia has now banned cultivation of wheat and is looking beyond its borders to acquire farmlands in other countries to ensure food security for its now much expanded population, or California is aggressively pursuing recycling of water. Another consequence of unlimited withdrawal of ground water is land subsidence which has become a dreaded occurrence in California and has caused enormous damage to highways, bridges, buildings and farmlands.

In order to meet their demands of fresh water, some of the water deficient countries are now promoting desalination of sea water by mimicking the natural process of rain formation through evaporation and condensation in solar green houses, or by accelerating the same process through formation of steam from sea water, and its subsequent condensation, in large engineering installations using industrial waste energy or solar energy. Even the medical procedure of dialysis, used in the treatment of chronic cases of malfunctioning kidney, has been scaled up to eliminate salt from sea water at economically competitive prices.

India has also been using ground water for irrigation, besides its use for domestic applications in many large cities and townships. The green revolution brought about by farmers in Punjab, Haryana and Western Uttar Pradesh was largely due to irrigation by ground water, as a consequence of which the water table in some areas has been continuously falling down at dangerous rates. In some parts of India, wells which meet the domestic water requirement of rural India are going dry. The situation is very bad in Tamil Nadu and critical in drought prone regions of Karnataka and Maharashtra.

Almost 60 per cent of our grain production comes from irrigated farms as compared to 80 per cent in China and only 20 per cent in USA. Water, becoming a scarce resource, we cannot afford to use it irrationally. Of late, we have begun to grow paddy in rain deficient Rajasthan on the strength of dammed waters of Chambal river or are cultivating another water guzzling crop, that of sugarcane, in drought-prone regions of Maharashtra on the strength of ground water, simply because sugar mills are built in the region. If the situation of irrational uses is not reversed or rectified soon enough, chronic water shortages could lead to anarchy in many parts of the country.

In order to face the imminent water scarcity, many countries are trying to reduce their national 'water prints' by adopting multi-pronged approaches, such as stopping cultivation of water intensive crops, using biotechnology to develop hybrid or genetically modified high yielding crops so as to reduce the water foot print of the crop, or adopting modern methods of farming or irrigation, etc. Some European nations have reduced their areas of wheat cultivation, which they now import from South America and to use their own water for economically more rewarding activities. Similarly, China, known to be suffering from chronic water shortages, is acquiring pig farms in the USA and closing its own, in order to reduce its national water foot print without cutting down its consumption of pork while importing pork from the USA. In effect, it amounts to import of processed water, now in the shape of pork. Since the water print of beef cattle is high, there are vociferous groups in the United States (where per capita beef consumption is highest in the world) who are advocating switch over to other non-vegetarian food, such as chicken meat so as to cut down their own national water foot print. Indeed by promoting rational use of water in society and in industry, USA has reduced its per capita consumption of water, without jeopardizing in any manner neither the quality of life of its people or industrial production. In this connection, the name of Israel must be mentioned as it has made the desert land bloom by according water scientific respectability.

The situation in India is not hopeless as it is in Saudi, California or Libya because our ground water supplies come from, what are called unconfined aquifers but those of Saudi Arabia from confined aquifers. Unconfined aquifers are rain fed but the others were formed through geological activities several thousand years ago, i.e. their water is thousands of years old. And even that impossible task would take thousands of years to replenish them with the scanty present day rainfall. And that could be possible only if the space vacated by pumped out water is not otherwise destroyed by subsidence of the overhanging land mass.

In India, we must ensure that every drop of rain water that falls percolates down into the ground and that its natural habitats in the Himalayas or Western Ghats, or the Terais and rivers or underground resources are not destroyed by human activity on the surface. The consequences of such wanton destruction manifest variously – the drying up of rivers and streams, flash floods even with mild rainfall, unpredictable monsoons, excessive rain in one go, etc.

Traditionally, Indians revere water as a manifestation of divinity and were the first amongst the nations of the world to take river waters to far flung areas and to harvest rain water through construction activities such as those which not only ensured their sustainability but exhibited their profound understanding of the geological formations and the flow of water in the terrains. Under the impact of all pervading western sciences and technology, Indians forgot their own knowledge-based glorious traditions and joined the race to 'unwrite' the laws of Nature. This led to widespread destruction of natural habitats of water in the Himalayas, the Western Ghats, river basins and underground.

Whilst the governments of India and of all the states, as well as some enlightened NGOs have from time to time launched serious efforts to rectify the rapidly deteriorating situation, their efforts can succeed only with cooperation of public motivated with scientific knowledge. This book addresses itself to the common man/woman so that he/she is able to comprehend various facets associated with the comprehensive personality of water. Water is the common concern of all Indians as it is truly secular.

Reminding them of their glorious traditions for ensuring abundant availability of water, the book seeks to familiarize them with the origins of fresh water from saline water which had exclusively covered its surface and created its habitats through geological activities extending over thousands of years. It explains the scientific principles of the nature designed fluvial processes which created the largest habitat of fresh underground water in the form of Sindhu-Ganga-Brahmaputra alluvial plains and transformed the Western Ghats into the 'water bucket' of southern India. The book provides an overview of global water resources and provides insights into the Indian scenario. And a Damocles' sword in the form of water scarcity is hanging over the Indian sub-continent. It seeks to introduce the emerging concepts of water foot prints of nations and commodities in order to promote rational utilization of water resources.

The underlying basic geological and scientific concepts are explained in simple scientific language so as to be intelligible to the 'common man' constituting the vast majority of Indians. His/her appreciation that all habitats of fresh water are creations of geology may help him to objectively evaluate the impact of any human activity (be it deforestation, mining, environment destroying chemical and other industries, excessive tapping of ground water, etc.) on the sustainable maintainability of habitats of water in their pristine glory. Thus, armed with scientific knowledge, they can develop ingenious methods to protect the ecological heritage, and yet have economic development of the country. No one has the right to destroy what Nature took millions of years to create. Empowered with knowledge, the emerging youthful India would intuitively protect the natural resources of fresh water, i.e. those available on the surface as well those underground, from abuse and over exploitation.

The book also seeks to prepare them to face the rapidly developing scenario of water shortages through conservation of naturally available renewable rains by water harvesting and avoiding wastage, including that by flood irrigation. Since the naturally available quantity of fresh water in the globe has remained almost the same over ages, and there is no escape from increasing demands on its availability, no other alternative than sea water desalination is available to augment the availability of fresh water. Anticipating that the country will necessarily have to adopt these techniques, scientific principles behind the three technologically feasible possible routes are explained in simple language. These are: (i) by mimicing nature in green houses, (ii) use of ultra-filtration techniques to remove salt from sea water and (iii) desalination through evaporation and condensation using waste thermal energy or renewable solar energy.

Indian efforts to bring about paradigm shift in creating fresh water reservoir within the sea – the Kalpsar Project – is mentioned so as to excite creativity in youthful minds, the exuberance of which signals the hope of future.

We are confident that armed with basic scientific knowledge, the innovative Indian mind would develop both new strategies and innovative technologies to harvest water without destroying Nature.

The Changing Scenario

It is pathetic that the modern man has relegated the august status of water to that of a commodity that can be relentlessly exploited for commerce and comfort. As a result of unregulated and uncontrolled use, many of its resources are already running low, some ground water resources are either exhausted or damaged beyond the possibilities of their rejuvenation. A measure of the extent of ground water extraction can be had from the fact that almost 6 per cent of global electric power generation is consumed for pumping out ground water, its treatment and its distribution. The ground water resource position is grim and their degradation seems to have resulted from lack of scientific knowledge about the nature of resources, their economic value and unscientific management. As population grows to an expected 9 billion by 2050, the already bad situation would turn worse. On one hand, more water would be needed for personal use, farming and industrial use, on the other, many ground water resources are getting exhausted; rainfall patterns are becoming erratic with climate change and surface topographical features that provide natural habitat for water are getting increasingly disturbed by anthropogenic activities. As a result, weather and climate-related extreme events, such as tornadoes, thunderstorms, storms, cyclones, floods and drought, now account for nearly 75 per cent of all natural disasters. They lead to an enormous toll of human suffering, loss of life and economic damage. Monitoring these events, predicting their movements and issuing timely warnings are essential to mitigate the disastrous impact of such events on population and economy. There is increasingly greater and mutually rewarding international cooperation in these areas of social concern.

A United Nations World Water Development report released in 2015 states that while reserves dwindle, the global water demand will increase by 55 per cent and that if the current usage trends do not change, the world would have only 60 per cent of the water it needs. The report states: 'Having less water available risks catastrophe on many fronts: crops could fail, ecosystems could break down, industries could collapse, disease and poverty could worsen and violent conflicts over access to water could become more frequent'. However, the silver lining is that the report foresees that efficient use of water could guarantee enough supplies in future. It calls on policy makers and communities to rethink water policies, urging more conservation as well as recycling of waste waters. The intelligentsia across the world is concerned with the man-made grim scenario of water scarcity and is concerned to create societal consciousness to reverse the situation and to promote responsible water use, conservation and development of water resources.

On the recommendations of the landmark 1992 United Nations Conference on Environment and Development (UNCED) in Rio de Janeiro, the United Nations General Assembly designated 22 March to be celebrated annually as the World Water Day. Commencing from 1993, each year, the day highlights a specific aspect of fresh water. All countries of the world celebrate the day as appropriate to their national context to promote public awareness through the production and dissemination of documentaries and the organization of conferences, round tables, seminars and expositions related to the conservation and development of water resources.

CHAPTER 2

Traditional Reverence for Water

In the social ethos of India, water has always enjoyed a position of reverence since the Puranic and Vedic times. It is an instrument of performing of all religious oblations, rituals and social events connected with purification or detoxification of the body and the soul. It is one of the five primal forces of Nature, called the *Panch Mahabhoots* or the five primary elements of life in the Vedic philosophy. These forces of Nature are: earth (*prithvi*), water (*jal*), air (*vayu*), fire (*agni*) and space or ether (*aakash*) and have been worshipped as gods since times immemorial. Harmonious interactions amongst them are responsible for the creation of life and for the sustenance of environmental ecology the planet earth. The Vedic philosophy stresses the importance of harmonious coexistence between human beings and natural resources. Protection of environment is articulated through informal rules in the form of prayers to respective gods of the five primal forces or the *Mahabhoots*.

India, area-wise, is the seventh largest country in the world, destined to become its most populous country by 2020, and with linguistic and religious diversity, it is united by its rivers and woven into same culture by their holy waters.

Scientists believe that organic life started from the unicellular organism called plankton in the sea and progressively evolved to its present day multifarious forms. This modern scientific understanding of evolution of life echoes what is stated in the Rig Veda that all forms of life on earth have evolved from *apah* (water); and water is a basic need of all living creatures of this earth. Thus, amongst the five forces of Nature, water is supreme. Vedic mythology expresses the same thought by stating that Lord Vishnu, also known as Narayana or the Supreme Lord of Universe rests on the cosmic couch formed by the coils of the great serpent of eternity in water while He waits for Brahma (the second of the Hindu Trinity) to recreate the universe *to recreate itself.* It is not, therefore, surprising that the first incarnation or *avatar* of Vishnu took the form of *matysa* or fish.

The Darwin's Theory of Evolution is more or less in conformity with the form of the evolution of life as reflected in the subsequent eight *avatars* (or the ten incarnations or the *dashavatars*, as they are called). The second *avatar* as *kurma* or tortoise suggesting that life started to move from water (sea) to land, followed by that of *varaha* or wild boar, i.e. life not with much intellect, which can survive on mushy lands.

The fourth took the form of Narasimha, or half man and half animal, suggestive of evolution from animals to intelligent beings. The fifth incarnation took the form of *vamana* or the midget or the *Homo sapiens* of the Darwin Theory. This was followed by that of Parashuram, or the man who lived in caves and wielded an axe for 'living'. Rama was the eighth incarnation, indicative of the evolution of life to societal levels. Rama combined in Himself the facets of family life, statesmanship, duty towards society and environment and introduced concepts of good governance wherein voice or feelings of all sections of the society are respected. The ninth, in the form of Krishna, represents the final quest for enlightenment and the yet to appear the tenth *avatar*, Kalki of the Man, will genetically be the most supreme in the universe.

It is well known that human life begins and is nurtured in its initial stages in the midst of amniotic fluid, mostly water, in the womb of the mother. In all the ancient Indian scriptures – four Vedas, Purans and Upanishads, water is seen as nurturing life; therefore, fit to be worshipped as God. The Vedic culture recognized this basic truth and the Rig Veda contains several *shlokas* or hymns in praise of 'water', extolling its virtues. Thus:

To the Waters, for Strength and Power

Ye, Waters, truly bring us bliss: so help ye us to strength and power,
That we may look on great delight.
Here grant to us a share of dew, that most auspicious dew of yours,
Like mothers in their longing love.
For you we fain would go to him to whose abode ye send us forth,
And, Waters, give us procreant strength.

I pray the Floods to send us balm, those who bear rule o'er precious things,
And have supreme control of men.

To the Waters, for Health and Wealth

The Waters be to us for drink, Goddesses, for our aid and bliss:
Let them stream health and wealth to us.
Within the Waters – Soma thus hath told me – dwell all balms that heal,
And Agni, he who blesseth all.
O Waters, teem with medicine to keep my body safe from harm,
So that I long may see the Sun.
The Waters bless us, all that rise in desert lands or marshy pools!
Bless us the Waters dug from earth, bless us the Waters brought in jars, bless us the Waters of the Rains!
<http://www.hinduwebsite.com/sacredscripts/hinduism/athar/avbook01.asp#hym04>

Or
'May the waters from the snowy mountains bring health and peace to all people? May the spring waters bring calmness to you; may the swift currents be pleasing to you; and may the rains be a source of tranquility to all ... and may the healing powers of water be available to all human beings. Water of river, well, pond, etc. if used and managed efficiently will reduce the intensity of drought and water scarcity. May the waters from the wells dug by humans be god to them, and may the healing powers of water be available to all human beings'.

Water is seen as an embodiment of self-awareness. It is referred to as a source of life, protector of earth and environment, nectar, honey, generator of prosperity and cleanser of sins. It is revered as pure, faultless, rain-shedding, sweet essence of the earth. In short, water is *divyajal* or simply divine. Water is regarded as the first important element that gave rise to the subsequent evolution of the universe.

The Vedic philosophy assigned each facet of the multi-dimensional functions of water to separate deities – Varun, Prajanya, and Indra – but their relative prominence changed from time to time. Varun ruled over all the waters in the different habitats (the seas, lakes and rivers as well as the underground), Prajanya ruled over clouds and is shown both as kind and cruel – kind when it rains gently and cruel when its lightening destroys crops and human habitats, including excessive rains or cloud bursts. Indra took over Prajanya when he destroyed the dark demonous clouds, personified as demon Vritasur, through his attack weapon, (called the *Vajra ne* Lightening), in order to compel the demon to shower rains over Bharat (India). The worship of Indra was demolished by Krishna Himself to be replaced by worship of elements of Nature as symbolized by Govardhan Hills. (Appendix A). Earlier, Rama had humbled Varuna (Appendix B).

Use of Water in Social and Religious Functions

The physical and aesthetic properties of water give it a unique mythical-religious niche. Therefore, it has played an important role in socio-religious and cultural traditions in India. Water occupies a unique status as no invocation is complete without water. Almost all religious rituals require sprinkling of divine water as an essential part of the ceremonies. The water is first purified with *mantras* to make it divine by invoking the presence of sacred rivers. Thus sanctified, the water is used for anointing the idols and then distributed to the devotees as *charanamrit*. On auspicious occasions like marriage, water filled pitchers, symbolizing rivers, are considered auspicious and kept as good omens at the marriage venue. Traditionally, all temples have big water tanks/ponds and wells in their vicinity where the devotees are expected to take a bath before entering the temple. On specific occasions or specific planetary configurations, a dip in the holy waters of the rivers is a part of many rituals.

Another form of expressing veneration for water is to organize religious and cultural fairs on the banks of rivers. The most sacred amongst them all is the Kumbh Mela. Millions of pilgrims take a dip in the holy river in order to wash away their sins and to cleanse their souls as they bathe. The Kumbh Mela takes place every three years, in turn, at the following four locations in India: (a) Prayag, at the confluence of rivers Ganga, Yamuna and mythical river, Saraswati; (b) Haridwar, where the River Ganga enters the plains from the Himalayas; (c) Ujjain, on the banks of the river Shipra; (d) Nasik, on the banks of the river Godavari. In addition, Ardh Kumbh Mela is held every six years at Prayag and Haridwar. *Purna* (complete) Kumbh Mela, the biggest and the most auspicious fair, is held every 12 years, always at Prayag. It is believed that at the time of churning of the seas, jointly by the gods and the demons, when a *kumbha* (an urn) containing *amrit* (nectar guaranteeing immortality) came out of the seas, the gods grabbed it and ran away with the urn, with the intention of denying its share to the demons. As they did so, a few drops of the nectar fell at each of these places, making these places holiest amongst the holy.

Verily, these *melas* are organized to express gratitude to the forces of nature responsible for good quality of life and prosperity. In the chain of such socio-religious functions, another important festival is that of *Chhath*, celebrated with great gusto in many parts of north India. This festival is dedicated to the god of Energy, i.e. the Sun. People seek the blessings of Lord Surya for life on earth in all its bounty by praying to Surya *bhagwan* (Sun God), standing in the holy waters of a river and carrying agricultural produce in straw baskets on their heads. The biggest gifts of solar energy to the earth are water and food. Hymns from Rig Veda in praise of Sun god are sung.

Makar Sankranti is yet another festival celebrated all over India and in many other countries of South Asia. It marks the transit of Sun into the zodiac sign of Capricorn (*Makar*) on its celestial path and heralds the beginning of auspicious times. As described in scriptures taking a bath in sacred rivers (Ganga, Yamuna, Sindhu, Saraswati, Narmada, Godavari and Kaveri) and giving alms on this day is considered as a good deed (*punya karma*). This day is particularly sacred because on this day, the celestial river, Ganga, on its first journey on earth, starting from the Himalayas met the sea at a place called Gangasagar in the Bay of Bengal. A dip in the holy waters of the confluence is considered as one that spiritually cleanses the soul and the body.

Other than Makar Sankrati, a few other days like Kartik Purnima, Magh Purnima, Ganga Dussehra, Mauni Amawasya are also regarded auspicious when it is equally fruitful to take bath during the zodiacally determined time period called the *Punya Kala* because during this period on this day, gods reside in water to listen to prayers.

It is very significant that the holy waters of any of the rivers make no distinction between men and treat the kings and powerful alike with the downtrodden.

Cultural Reverence for Rivers in India

Water and human life are so closely inter-related that our social behaviour is instinctively defined by water. India is the only country in the world where folklore accords divine aura to rivers and mystical stories are woven around them. All the major rivers are blessed with divine or celestial origins – whether they be Ganga,

Kaveri, Yamuna, Gomati, Godavari, Narmada, Tapti, Chambal or the demon-like Damodar and Kosi – the sorrow of Bihar, or the lake Mansarovar, across the border. In this context, the holy river Ganga is regarded as truly celestial and divine. The other rivers in different parts of the country are her manifestation only. Mythologically, Ganga flows with equal ease in the three *Loks* (*Dev Lok* – Heaven, *Bhoo Lok* – earth and *Paataal Lok* – the Nether world). Therefore, Ganga is often addressed as *tripathgaa and trisota* (having three paths and three different sources of origin). In the 10th chapter of Shree Bhagavat Gita, Shree Krishna honours Ganga by saying that Ganga is His own reflection. In mythology, Ganga emerged from the *Brahmanda* (Universe) when the upper shell of the Brahmanda got pierced with the big toe of His left foot in His action in the *Vaman* avatar to measure the Universe in order to wash His lotus feet. As the water had directly touched the feet of Vishnu, it became *Vishnupadi* (*pad* means foot) and remover of the sins of the entire world. After thousands of *yugas*, She descended to the *Dhruvalok* (the Polar Star), then to the Saptarishis (constellation of Seven Great Sages or the Great Bear). The Saptarishis carry the celestial Ganga with great reverence on their matted hairs. From there, following celestial passageway, the holy waters descended to Mount Kailash on the Himalayas in the *Bhoolok* (the earth) and got entangled in the locks of Shiva. And Ganga got another name of *Jata Shankari.*

As Shiva let loose only a trickle of her mighty waters, out from His matted locks (may now be scientifically regarded as the matted roots of Deodar trees holding soil soaked with rain water), the waters rushed down the Himalayan slopes with such ferocious velocity as destroyed all that came across their way until they reached the hermitage of Sage Jahnu who swallowed her gushing waters completely (reminding of the similar action performed by the Bhabhar zone at the foothills of Himalayas and discussed later in a chapter); but soon released it, taking compassion on the pleadings of Bhagiratha, so that the holy waters of Ganga may breathe life afresh in otherwise lifeless bodies of his ancestors. Ganga, thus, got a new name – that of Janhvee, i.e. the daughter of Jahnu. It is in this name that the Adi Shankaracharya addressed the Ganga in his famous *Ganga Strotam,* composed in Sanskrit.

In peninsular India, the Godavari river is often called The Ganga of the South or even *Budhee Ganga* (or aged Ganga suggestive or reflective of the geological understanding that the river Godavari was existing even when the Ganga had not descended from the Devlok to the Bhoo Lok. Mythology has it that in order to redeem Sage Gautam whose hermitage was near Nasik, Lord Shiva let another stream of Ganga waters emerge out from his matted locks at Nasik, as if to remind that waters of all Indian rivers are as holy as that of Ganga herself. Similarly, if the mighty Ganga could be accommodated into the small wooden vessel *(kathauta)* of the tribal King Kevat so that he could wash the feet of Shri Ram, so were the waters of the river Kaveri accommodated in the *Kamandal* of Sage Aagustya.

Hymns sung in praise of Ganga abound, from Vedas and Puranic stories to those by Adi Shankaracharya, who composed the famous 'Ganga Strotam' and Jagannath Pandit's 'Ganga Lahiri', in more recent times, 1592–1666 CE. The english translation of Ganga Strotam is given below:

> O Goddess Ganga! You are the divine river from heaven, you are the saviour of all the three worlds, you are pure and restless, you adorn Lord Shiva's head. O Mother! may my mind always rest at your lotus feet.
>
> O Mother Bhagirathi! You give happiness to everyone. The significance of your holy waters is sung in the Vedas. I am ignorant and am not capable to comprehend your importance. O Devi! you are full of mercy.
> Please protect me.
>
> O Devi! Your waters are as sacred as 'charanamriti' of Sri Hari. Your waves are white like snow, moon and pearls. Please wash away all my sins and help me cross this ocean of Samsara.
>
> O Mother! Those who partake of your pure waters, definitely attain the highest state. O Mother Ganga!
> Yama, the Lord of death cannot harm your devotees.
>
> O Jahnavi! Your waters flowing through the Himalayas make you even more beautiful. You are Bhishma's mother and sage Jahnu's daughter. You are saviour of the people fallen from their path, and so you are revered in all three worlds.
>
> O Mother! You fulfil all the desires of the ones devoted to you. Those who bow down to you do not have to grieve. O Ganga! You are restless to merge with the ocean, just like a young lady anxious to meet her beloved.

O Mother! Those who bathe in your waters do not have to take birth again. O Jahnavi! You are held in the highest esteem. You destroy your devotee's sins and save them from hell.

O Jahnavi! You are full of compassion. You purify your devotees with your holy waters. Your feet are adorned with the gems of Indra's crown. Those who seek refuge in you are blessed with happiness.

O Bhagavati! Take away my diseases, sorrows, difficulties, sins and wrong attitudes. You are the essence of the three worlds and you are like a necklace around the Earth. O Devi! You alone are my refuge in this Samsara.

O Ganga! Those who seek happiness worship you. You are the source of happiness for Alkapuri and source of eternal bliss. Those who reside on your banks are as privileged as those living in Vaikunta.

O Devi ! It is better to live in your waters as turtle or fish, or live on your banks as poor 'candal' rather than to live away from you as a wealthy king.

O Goddess of Universe! You purify us. O daughter of muni Jahnu! one who recites this Ganga Stotram everyday, definitely achieves success.

Ganga *Aarti*

So intense is the Indian veneration for the Ganga that every morning and evening *aarti* is performed of the river as an expression of gratitude and for invoking Her blessings. Often called the 'ceremony of light', the *aarti* involves waving lighted wicks ceremoniously on the banks of the holy river. It is a sight to see and the ceremony has become an attraction for the tourists.

References

Hymns of Atharvaveda: Hinduwebsite.com: Translation by Ralph Griffith, http://www.hinduwebsite.com/sacredscripts/hinduism/athar/avbook01.asp#hym04 (retrieved January 2017)

Important Vedic Quotations on Rivers & Water, https://tamilandvedas.com/2014/05/08/important-vedic-quotations-on-rivers-water/ (retrieved January 2017).

Sharma, K.N. Environmental Protection and Water Reverence in Vedic Culture, ICID, https://www.scribd.com/doc/16813661/ (retrieved January 2017).

Shocking Secrets of the Vedic Science Revealed!, http://mcknowledge.info/about-vedas/what-is-vedic-science/ (retrieved January 2017).

Vedic Hydrology, nihroorkee.gov.in/rbis/vedic.htm (retrieved January 2017).

Water: The Divine Elixir of Life – Indian Scriptures, http://www.indian scriptures.com/vedic-lifestyle/customs-and-traditions/nature-worship/water-the-divine-elixir-of-life (retrieved January 2017).

Appendix A

Krishna's foster mother Yashoda is preparing for the puja for the worship of Lord Indra.

'But why worship him?' Krishna asks.

'Don't say that my child!' Yashoda warns. 'His wrath can be terrible.'

'Why worship such a terrible god?' Krishna asks. 'It is not good to get angry. A god who gets angry if you don't worship him is no god. Puja (to worship) is divine, but no god can compel you to worship him. There will be no worship of Indra.' Krishna advises to worship the worthy.

'And who is worthy of worship?' his foster father asks.

'We are cowherds, father,' Krishna tells him. 'Our life is linked to the cow. So, we should worship the cow. She gives us milk and butter. So, worship the cow, mother. The cow, in turns, lives on the forest. If there were no forests, where would the cows graze? So, mother, worship the forest. If there were no forests, there would be no cows. Worship the Yamuna which waters the trees. Worship the mountains against which the clouds burst and because of which the forests remain green.'

'Who gives us medicine in the form of magical herbs and plants? Who sends us clean water and air from the top of its peak? And who gives us good grass for our cows, so that they give us milk that's sweeter than honey? It is Govardhan!'

'Worship these, mother. They are worthy of worship, as is the Govardhan mountain. Do not fear Indra's wrath. Only he is worthy of worship who does not get angry. You are worthy of worship since you never get angry.'

At hearing this, Indra becomes enraged and sends a terrible storm. Krishna calmly stands outside playing his flute while thunder and lightning rage all around. The villagers become frightened and approach Krishna, telling him that he'd better ask forgiveness or Indra will destroy everything. Krishna just smiles, which upsets them even more. 'Do not fuel Indra's wrath. Didn't you prevent his worship? Now ask the Govardhan to protect us!'

Smiling, he lifts the Govardhan mountain and they all seek shelter beneath it from the storm. At this point Indra, humbled, begs forgiveness.

Krishna was advising that we should worship Nature herself and in particular those local parts of nature on which we ourselves depend. *Krishna was indeed promoting bioregional animism, which, for the cowherds in his village was centred on cows*

Adapted from: <https://mistressofthehearth.wordpress.com/2013/05/16/krishna-orders-the->.

Appendix B

Faced with the dilemma of how to cross the ocean to Lanka, where his abducted wife Sita is held captive by the demon king Ravana, Rama performs a penance (*tapasya*) to Varuna, the Lord of Oceans, fasting and meditating in perfect *dhyana* for three days and three nights. Varuna does not respond, and Rama arises on the fourth morning, enraged by the god's arrogance. With his bow and arrow, he angrily begins attacking the oceans with celestial weapons – burning up the waters and killing its life and creatures. The *vanaras* (monkeys) are dazzled and fearful at witnessing the enraged Rama demolish the oceans, and his brother, *Lakshmana*, prays to calm Rama's mind.

Just as Rama invokes the *brahmastra*, considered the most powerful weapon capable of destroying all creation, Varuna arises out of the oceans. He bows to Rama, explaining that he himself was at a loss to answer Rama's question. Begging him not to destroy the oceans with the missile, he suggests that Rama re-direct the weapon at a demonic race that lives in the heart of the ocean. Rama's arrows destroys the demons, and establishes a purer, liberated environment there.

Varuna promises that he would keep the oceans still for all of Rama's army to pass, and Nala constructs a bridge (*Rama's Bridge*) across to Lanka. Rama justifies his angry assault on the oceans as he followed the correct process of petitioning and worshipping Varuna, but obtaining the result by force for the greater good.

From, the Ramayana <https://en.wikipedia.org/wiki/Varuna>.

CHAPTER 3

Origin of Fresh Water on Earth and its Habitats

According to both Indian and Greek mythologies, there was only saline water on the surface of primordial planet earth. Organic life first appeared in marine waters. Both the mythologies converge on the belief that water is subsequent to none for hosting or sustaining life. Geological scientists of modern times have provided unique confirmation to the beliefs of ancient mythologies through their studies of fossils. Their studies show that the oldest fossils found are those of marine life. There was no trace of salt-free (the so-called fresh water) water then as the available land surface was not suitable to provide space where fresh water from rains could make its distinctive habitat/s. The primordial land surfaces then available were rocky and truly solid. Lands with mountains and valleys and top soil, with which we are familiar today, were non-existent then and appeared much later.

Earth is the only planet of Sun, so uniquely placed in the solar system and blessed with atmosphere that it could have just the right climate to sustain the eternity of liquid water, including fresh water, so as to be able to support human and other forms of fauna and flora in the form we know. Of them, the human form is wholly dependent on fresh water. In retrospect, it appears that the plans of the Creator went awry as He wanted to create mankind as an image of Himself but soon found that the earth had only saline water which cannot sustain mankind, though it could sustain marine life. The Nature had then to make gigantic efforts to create fresh water and its habitats on earth, away from the sea. These efforts invoked severe geological activities over millions of years. During all these long years of geological evolution of its anatomy, its surface topography was being shaped by the phenomena of weathering and erosion with the help of the primary elements of Nature – running water, wind and the Sun – i.e. three of the five *Mahabhoots* of Indian mythology.

Fresh water is a unique gift of Nature to us on Earth. It is not known to be available anywhere else in the universe in as much abundance as on earth; above all in a temperature range in which organic life is sustainable. It is unique on other counts as well. It cannot be destroyed or created though it may change its form. This fact imparts eternity and divinity to water. Despite passage of millennium on millennium years, we still have it in the same quantity as was available in the very beginning of life on Earth. The same quantity has been continuously recycling itself and must continue to do so till eternity, if mankind is to survive on earth. In order that water is available to us till eternity, we must zealously guard its natural habitats, both on the surface and underground as well as those at the polar ice caps or up in the mountains and in the atmosphere. We must ensure that the naturally occurring process of its recycling continues to occur in the manner designed by Nature, without any hindrance.

Natural Habitats of Fresh Water

The natural habitats for fresh water are:

- As ice on polar caps and at high altitudes of mountains, glaciers and as permafrost;
- In rivers and lakes (including ponds) on surface of the earth;
- Underground as soil moisture, also as saturating/filling the spaces available in the alluvial aggregate of soil particles; this is the source of water in the traditionally dug wells;
- As aquifers in saturated soil and in fragmented rocks; the fragmented rocks may be located deep underground or may be in the rocks up in the uplands and mountains;
- Importantly, also as moisture in air, clouds and rains; and
- In human bodies or in insects, animals and plants and plant products – wheat, sugarcane, fruits, etc.

Water is not static in any of the habitats but continuously changes position from one habitat to another though its habitation time in the different habitats greatly varies with the shortest being in the atmosphere and longest in deep aquifers and in depths of the ocean.

Fresh water plays a central role not only for sustaining life but also in shaping all societal activities of mankind. According to astronomical verification of certain cosmic events described in the Indian epic Mahabharata, civilized human existence is perhaps about 10,000 years old. During all these long years, the sustainability of human life was never in doubt as water was available in abundance and the sanctity of the natural resources of water was not challenged by human activities. Today, however, the position is not rosy. Dark clouds are appearing on the horizon, threatening the very survival of human beings on earth as anthropogenic activities have violated the natural resources of water. The position has to be reversed at once to avoid occurrence of catastrophic happenings in the world.

For our efforts to succeed in this direction, we must know how Nature had crafted the habitats of fresh water. Our attempt in this chapter of the book is to provide a glimpse of the gigantic geological processes which occurred over millions of years ago to create them. Nature also provided a cover of vegetation to either protect the habitats from the very forces (weathering and erosion) that created them or to accentuate the processes of collection and retention of fresh water in their habitats in their pristine sanctity. Trees help our soil to remain healthy by reducing soil erosion and by creating a soil climate suitable for micro organisms to grow.

Earth in its Primordial Form

There is universal agreement amongst scientists that at its very beginning as a planet of the Sun, the Earth was an almost spherical huge ball of fire, about 12,800 km in diameter but slightly flattened at the poles. Its interiors were (and still are) at increasingly higher temperatures at greater depths. Over a period of time, measured on geologic scale, it cooled down. At this, its almost nascent state, the global sphere could be divided into a few geo-physically distinct but nearly concentric temperature zones. Starting from the centre and going outwards these are:

- The inner core, in enforced solid state and consisting of mostly of iron and nickel, the outer core, also consisting of iron and nickel but in liquid state.
- The mantle, made up of semi-solid mixture of various minerals or materials. It is made up of three layers: the deepest being the lower mantle in semi-solid state; the outer mantle and the asthenosphere which is a partially molten zone, and finally the crust. The upper mantle and the crust together constitute the asthenosphere.

The temperature, however, increases as one gets deeper into the Earth. The Earth's core is estimated to have a temperature of around 5,000°C. The temperatures at the crust's interface with the outer mantle range between 500°C to 900°C, and to over 4,000°C at the interface of the lower mantle with the core.

It may be mentioned for the sake of scientific clarity that, despite the high temperatures, the core, consisting mostly of iron and nickel, is fully in solid state. Although most metals are in molten/vaporous state at such high temperatures at atmospheric pressures, their occurrence in solid state in the mantle is solely due to the physical phenomenon whereby the melting temperatures of minerals or rocks or metals are raised by increasing the pressures over them. In this case, the enormous weight of the lithosphere and the mantle over the core exerts enormous pressures on the matter inside. In other words, in the event of the pressure from the over burden being catastrophically removed (and this can happen only through geological activity), the matter in the induced solid state would at once turn into molten or vaporous state. This scientific truth often reveals itself on volcanic eruptions when molten lava flows out along with dust and gases. The spread of the plume of fine dust in the atmosphere for several kilometre around the volcano when it erupts is also a scientifically observed phenomenon. The plume may contain very fine dust and its spread far and wide in the atmosphere is known to disrupt air traffic in modern times.

The Lithosphere

The lithosphere (crust and the uppermost solid layer of the mantle) is the coolest of the Earth's thermal layers. It is the most significant part of the Earth's mass that has decided the physical and geomorphological features of its surface. Its density is the lowest as the heavier matters of the Earth (e.g. iron and nickel) settled down inwards the core under the influence of gravitational forces. Chemically, it is made up of silicate minerals, sodium, and potassium with iron, calcium and magnesium. In geological terminology, it is made from three different types of rocks which the geologists call the igneous, the sedimentary and the metamorphic rocks.

The lithosphere is made by the spurting out of the magma or molten mass from the underneath from the crust of the earth on to the surface either through massive and molten expulsions through volcanic eruption/s (called extrusion) in the form of lava or when the spurting out lava is pushed into the pre-existing fissures in the adjacent crust/rocks (now called intrusion) and formed dykes (or dikes) within the pre-existing rocks. The rocks formed on solidification are called igneous rocks (magma is a mixture of molten or semi-molten rock, volatiles and may also contain suspended crystals of minerals, dissolved gases and bubbles).

Igneous rocks come in two varieties, called granites and basalts which formed from the cooling of magma. They are different from one another predominantly because of two factors: (1) they may have cooled at different rates and (2) their 'mother' magma (original melted rock) was of a different composition. When the magma cools at different rates, it creates different sized minerals. Quick cooling magmas, called basalts, are fine grained and their mineral distribution can only be seen under microscope. Basalts often form on sea bottom from volcanic 'extrusive' eruptions the magma. Because the magma comes out into the sea bottom, it solidifies rapidly. Therefore, it has a fine texture. Extrusive rocks may rise close to the surface where they are subject to weathering processes.

Granite is the most common rock type on the continental land masses and granitic 'basement rock' can be found just about everywhere below the surface. Granite is intrusive, which means that the magma was trapped deep in the crust, and probably took a very long time to cool down, enough to crystallize into solid rock. Increased time taken to solidify allowed the separating minerals to grow, resulting in the formation of coarse-textured rock in which individual mineral grains are easily visible.

Both, basalts and granites are made up of silicate minerals of sodium, potassium, magnesium, iron and calcium. Their similarity ends here. Basalts have relatively lower silicon, sodium and potassium and higher amounts of magnesium, iron and calcium whereas granites have higher amounts of silicon, sodium and potassium and lower amounts of iron, magnesium and calcium. This difference makes the granite rocks lighter than the basalts. Since both types of solidified rocks are floating on crust, basalts sink deeper into the crust than granites. Consequently, a relief pattern is formed on the surface of the earth, with water filling up the 'deeps', forming oceans and the 'ups' forming continental land masses.

Sedimentary rocks are formed by an initial weathering of igneous rock. The weathering is usually associated with water and wind which facilitate erosion, transportation, deposition, and cementation over long period of time. The fine weathered particles, pebbles and boulders are carried by flowing waters in rivers and deposited into sedimentary basins, provided by ancient lakes and seas (e.g. The Tethys Sea), forming layers. High pressures slowly convert these layers into sedimentary rocks. The Himalayas are good examples of sedimentary mountains.

Since the volcanic eruptions occurred repeatedly, new lava spread on the pre-existing solidified lava spreads (from the preceding flows). The heat (and pressure/weight from the overburden of existing igneous rocks) brought about both chemical and physical alterations. The resulting rocks are called metamorphic rocks. These forces also caused the development of cracks and voids in what was earlier a massive solid block. Examples of metamorphic rocks are schist (converted basalt), gneiss (converted granite) quartzite (compressed sandstone), and marble (compressed limestone or dolomite). These are denser than the parent rocks and much less susceptible to disintegration by erosive forces. Metamorphic rocks are, therefore, those igneous or sedimentary rocks that changed their physical and/or mineralogical character by great pressures and temperatures within the crust and upper mantle of the Earth. The high pressures not only fragmented the solid rocks into pieces, but also brought about changes in the chemical make-up of the rock.

All three types of rock make up the earth's lithosphere. Sedimentary rocks are the most abundant rocks only on the surface of the Earth, but igneous and metamorphic are abundant deeper into the mantle.

The lithosphere is solid and strong enough to support the weight of oceans and continents. The earth has two types of lithospheres: the portions that support the oceans are called the 'oceanic' and the other 'continental' lithosphere. Oceanic lithosphere is the 'rock bottom' of all oceans of saline water. The continental crust is about 20–70 km thick and the oceanic crust about 5–10 km thick. The latter got overlaid by oceans after it had selectively cooled down sufficiently enough to permit water to remain in liquid state. The crust may be regarded as floating on mantle and the relief of its outer surface is uneven. The environmentally induced weathering of rocks gave rise to the formation of soil, including the alluvial soil. Truly, it is the soil that supports humans, their agriculture and many of their resources of fresh water.

The rocky mass of the earth has been formed by massive repeated volcanic eruptions of lava from the mantle which on emerging on the surface was instantly decompressed and turned into molten lava, spread and covered large areas of the surface. Since the chemical or mineralogical composition of the mantle is not uniform, the character of the extruding lava which formed the earthy rocks, also varied from time to time and from place to place.

We are told by geophysicists that new oceanic lithosphere is constantly being produced at mid ocean ridges by conductive cooling which converts the hot asthenosphere into lithospheric mantle. This makes it obvious that interiors of the earth are not thermally static but their matter is subject to movement in the manner of convection currents in liquids. In other words, the internal earthly matter is under intensely agitated state due to geophysical forces.

Lithosphere and the Plate Tectonic Theory

The lithosphere, i.e. the outer rigid layer of the earth is now being increasingly called by geophysical scientists as tectonic 'plates'. Geo-scientists believe that 60–250 million years ago, there were only two continental lithosphere plates: those of Gondwana in South and Angaraland/Laurasia in North. They were separated from one another with a huge spread of saline sea which was named as the Tethys Sea in 1893 CE, after the name of the wife of Oceanus – the God of all Seas in Greek mythology, akin to Varun in Indian mythology. Only the Greek and Indian mythologies have specific 'gods' for oceans.

These two tectonic plates broke into a number of plates – some large and some small – averaging about 80 km in thickness; they are shown in Figure 1. They host both oceans and continents. As the plates are resting on the soft underlying aesthenosphere, they are in a state of drift in respect of their positioning on the global sphere. They are not stationary but move with extremely slow speeds when compared with our speeds of travel or flow of water in rivers. Yet the slow push of a drifting plate to the relatively stable or stationary plate builds up large stresses at their interface. As a consequence of build up of stresses in the strata, tectonic activity is set in motion at the boundaries of the interacting tectonic plates. The activity is of titanic dimensions and may manifest itself in the form of most catastrophic geological events, gradually creating folded mountains, generating earthquakes and causing volcanic eruptions and even disappearance of seas. The Indian plate is drifting towards the Eurasian Plate and this region continues to experience earthquakes of varying magnitudes from time to time.

Figure 1

Tectonic plates

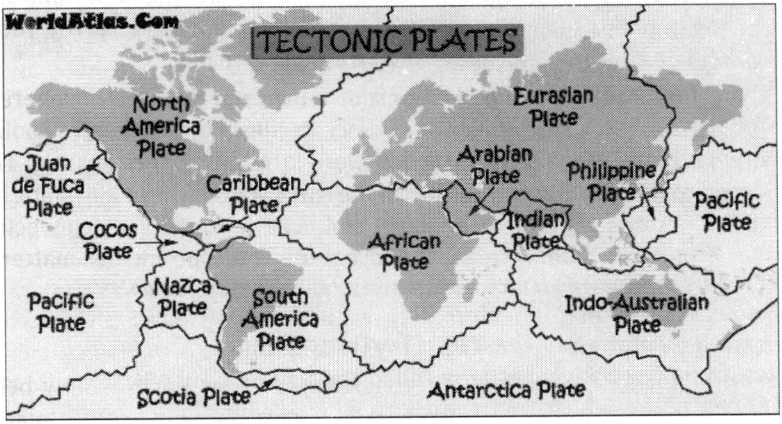

What physical forces are responsible for the movement of tectonic plates is a matter of geo-physical studies. It is held by some scientists that rotation of Earth on its own axis, shifts in geo-magnetic poles, local differences in densities of matter in the womb of the Earth, etc. may be having a prominent role in influencing them. Readjustments among the internal forces result in physical and violent movements in the crust and lithosphere of the Earth. It was through such violent movements that the present day surface topographical features of the Earth, including the habitats of fresh water, were developed.

An important point emerges from the difference in the mineralogical characteristics of the two types of lithospheres: the oceanic lithosphere is heavier or denser than the continental one. This fact has played a key role in shaping the topography of the world, including raising the Tibetan plateau to its present heights and creation of the highest mountains in the world. These mountains, their valleys and plateaus are the highest abode of fresh waters in liquid or solid state in pristine quality.

Indian mythology has blessed them with divinity; the most sacred amongst them being the Kailash Parvat, abode of Lord Shiva Himself, and the Mansarovar Lake in Tibet. The fact that these two places are located in two different countries recognizes in very subtle manner the fact of the universality of water.

From our point of view, tectonic movements helped to create habitats of water on the land mass, notably the emergence of the Himalayas, fragmentation and mashing up of its massive rocks and of the Deccan Plateau of peninsular India and shaping the topographical contours of India – including the Western and Eastern Ghats, and giving direction to the flow of monsoon winds and rivers, etc.

As all this was happening, the Sindhu-Ganga-Brahmaputra alluvial plain (itself a huge storehouse of fresh water) was being slowly and gradually created by the settling down of the particulate matter created through weathering and erosion of the rock strata of the Himalayas and brought/transported down by the turbulent streams originating from the mountains. The Tethys Sea was simultaneously shrinking and ultimately vanished away totally. The vast emptiness of the deeps of the Tethys Sea, in course of geologic times, was replaced by what is now called the Sindhu-Ganga-Brahmaputra alluvial plains which are the thickest alluvial plains in the world.

State of Violent Unrest in the Bowels of the Earth

It may now be re-stated that the thickness of neither the tectonic plates nor of the crust underneath them is uniform over their entire spreads. They vary from place to place. This is also true in respect of their mineralogical composition and physiognomy including porosity, density or physical properties. Despite these inhomogeneities on massive scale, the tectonic plates maintain physical unity amongst them. They make up a single unbroken heterogeneous massive body by forcefully silencing the grumbling internal stresses or dissensions; apparently suggesting that there is tranquility among them by forcefully suppressing the discord among them. This situation is fully in accord with the basic precepts of mechanical sciences. The compulsions of maintaining physical unity in the heterogeneous body, introduces complex stresses in the body; at places the stresses are tensile in nature, at others compressive and yet at other places shearing. The rigidity of the frame of their capture keeps them docile and submissive. Engineers express this situation by stating that a dynamic equilibrium exists between the mutually hostile forces. If any individual member of the complex stress system is disturbed either by natural or extraneous human activities, these internal stresses or dissensions seek fresh alignments to strike a truce among them (or readjust themselves to strike a fresh equilibrium). Nevertheless, the one basic character of truce is its volatility. What is true for human behaviour is true geologically too.

Even human activities on the surface of the earth can disturb the peaceful accord among the elements of the strata. Many environmentalists apprehend that construction of dams to build massive stores for fresh water, otherwise going waste by the river carrying it to the sea, hold the potential to trigger seismic activity. Conversely, similar situation may arise on the development of huge hollows underground on account of pumping out of underground water from aquifers to near exhaustion. Minor seismic activity around the Koyna Dam, built on the geologically stable Deccan plateau has often been attributed to the Dam. Similarly, excessive mining activities or tapping of very deep petroleum deposits for water or natural gas hold real threat for triggering seismic activity. Indeed, the scientifically monitored land subsidences in parts of California (USA) lend considerable support to and provide unique confirmation to the apprehensions of environmentalists. Some associated facets are discussed later in the book.

Manifestations of Seismic Activity in Modern Times and Internal Thermal Energy

Thus, the outward manifestation of new internal alignments or adjustments is seismic activity, including earthquakes and volcanic eruptions – some very violent and some minor. Violent shake-ups of the lithosphere have created the habitats of fresh water. Whilst minor eruptions from active volcanoes are well known, some violent ones are also known; the worst in recorded history occurred on 27 August 1883 CE, when almost two-third of the island of Krakatoa (in Indonesia) disappeared in a single explosive and violent volcanic eruption followed by tsunami of gigantic dimensions. Together, they killed 36,417 persons. The catastrophic event was preceded by continued seismic activity both after, and prior, to the event. Science and scientific techniques were not as advanced then as they are today to have forewarned about the impending tragedy.

In more recent times, a new island was created by a volcanic eruption by an underwater volcano in Tonga in Pacific Ocean on 21 January 2015 but volcanologists are apprehensive that the ocean may eventually erode it away completely. Earlier, about 3,500 years ago, there was a massive eruption of Thera volcano in Greece. It is now recognized that environmental effects of Thera eruption were felt across the globe as far away as China and, perhaps, as far away as Antarctica and North America.

It is not always that Nature makes us aware of thermal energy in its bowels through violent outbursts only. It has been making us aware about thermal energy in its bowels through hot springs, which were used to warm up Roman baths. Elsewhere, in United Kingdom, a hot spring is credited to possess curative properties for skin ailments. In India, an aura of divinity is created around a few hot springs. In modern times, science and technology has advanced to a stage where their thermal energy is being tapped to produce electricity. Power generation plants based on geothermal energy have or are being established in India in Chhattisgarh, West Bengal and Gujarat.

Emergence of the Himalayas as the Single Largest Source of Fresh Water

Even the ordinarily invisible or insignificant movement of the smaller Indian tectonic plate resulted eventually with its collision with the massive Eurasian plate. The striking together of two tectonic plates induced movements of the rock strata of both the tectonic plates on massive scale. As a result, the geologically young strata of sedimentary rocks, formed from the sediments that had settled on the floor of Tethys Sea, got pushed up through formation of folds, forming synclines and anticlines; whereas its older portion - the Deccan Plateau - got mashed up, metamorphosed and geological faults and cracks occurred at many places to an extent that even extensive rift valleys were formed. These structural changes in the petrologic morphology made it possible for the Deccan plateau to host fresh water sources.

Significant geomorphologic changes also occurred on the Eurasian plate as well in accordance with the universality of the laws of Nature. Even before the Indian landmass physically collided with the Eurasian Plate, the Tethys Sea was getting subducted beneath the Eurasian Plate which was raising the height of the Tibetan plateau. The raising of the Tibetan plateau was accentuated when the Indian plate struck the Eurasian plate. The heavier oceanic lithosphere penetrated underneath the physically more massive continental lithosphere of the Eurasian plate. This is shown in Figure 2. This process raised the height of the Tibetan Plateau to over 5 km above the sea level. As the Indian subcontinent continues its ramming into the Eurasian plate, the Tibetan plateau continues to rise at a rate of about 5 mm every year. The Tibetan plateau serves as the head waters of the drainage basins of some of the largest rivers - Salween, Mekong, Yangtze and others.

As the Tibetan plateau rose up, the Himalayan and Karakoram mountain ranges were also formed. As they rose, the Tethys Sea

Figure 2

Interpenetrating ancient continental crust or plate through continental lithosphere

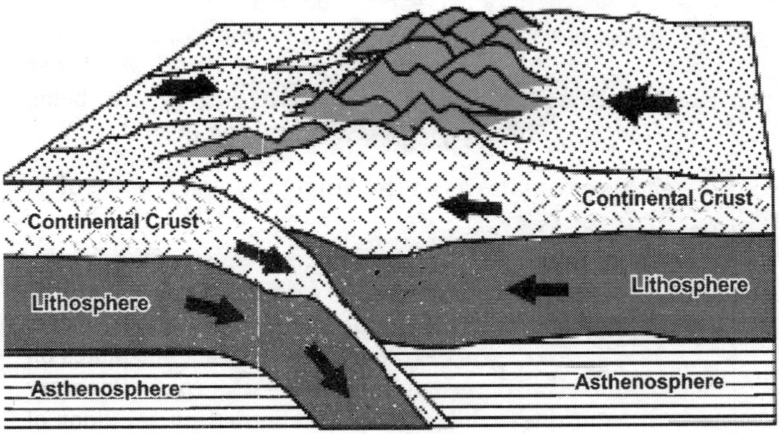

Continent – Continent Convergence

Source: Stephen A. Nelson. Tulane, University.

correspondingly contracted and ultimately vanished altogether. This is schematically illustrated in Figure 3. The Himalayan ranges are geologically called the young folded mountains.

The character of folded mountains can be understood with reference to Figure 4. In the absence of free space to move, the 'push' exerted by the drift of the Indian plate on the massive Eurasian plate, caused sedimentary strata of the Tethys Sea to get folded up, much in the same manner as thick carpet (spread on the floor) would do at its interface, when pushed against a rigid wall. It may be seen that all the sedimentary strata in Figure 4 do not show the anticlines. Truly, the anticlines were there when freshly formed but they got weathered and eroded by glaciers, water and wind. The eroded anticlines have turned into their outcrops on the surface. The eroded material was transported by water to long distances to fill up the void of the disappearing Tethys Sea.

The Himalayas are the largest, the highest (6,000 – >8,000 m) and the youngest folded mountains in the world and are stated to be still rising as a consequence of the continued drift of the Indian plate putting pressures on the Eurasian plate. At some of these heights, the

Figure 3

Rise of the folded mountain ranges of the Himalayas and the gradual disappearance of the Tethys Sea; the Gondwana is now called the India Plate

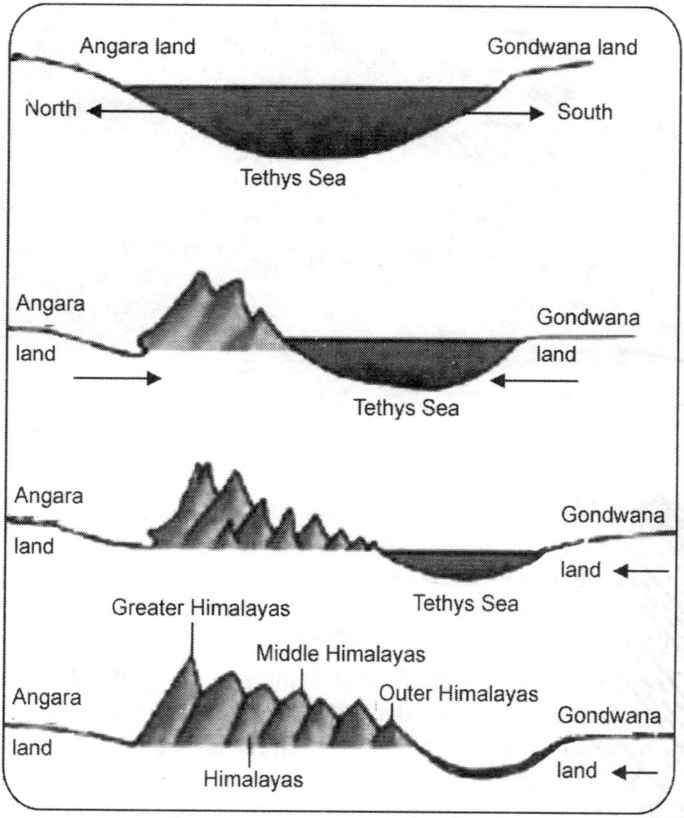

Source: Social Science Class X Textbook, Government of Tamil Nadu.

atmospheric temperatures are sub-zero all the year round. This means that the mountain tops are covered with snow and ice all the year round; the word 'Himalaya' of Hindi origin indeed means the abode of snow. The Himalayas are the source of many perennial rivers of northern India, some originating from the glaciers.

The Himalayan ranges are spread over a vast area; west to east, they are spread over a length of about 2,400 km; in the west their width is over 400 km which narrow down to 150 km in the east in Arunachal Pradesh. There are three principal ranges of mountains which together constitute the Himalayan ranges. These are:

Figure 4

Formation (schematical) of syncline and anticline on folding of sedimentary rock bed. the outcropping of some of the sedimentary beds as a consequence of erosion and weathering is also shown

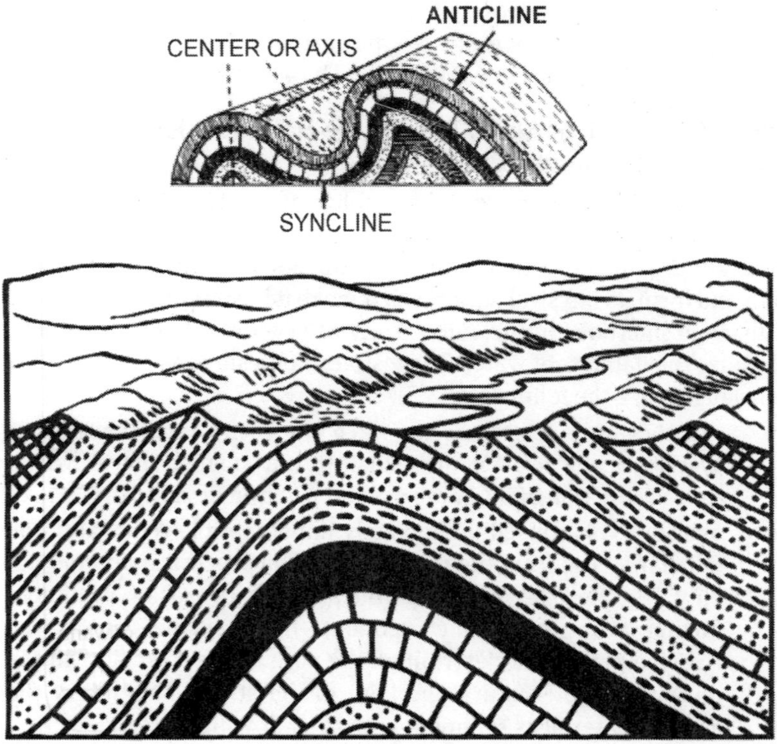

Source: https://en.wikipedia.org/wiki/AnticlineFormation_processes

- The highest and northern most called the Himadri mountain range; also called the Greater Himalayas. The higher regions remain ice bound all the year round and contain many glaciers. Whilst moving downhill into the valleys, they badly scour the surface and carry with them the thus generated rock detritus.
- Running almost parallel to the Himadris is the Himalayan range with heights in the range 3,700–4,500 m and width of about 50 km.
- The southernmost mountain range of the Himalayas is the Shivalik range, of heights as low as 100 m to as high as 1,000 m.

Together, the three ranges constitute a unique naturally designed system to arrest the north bound, water laden monsoon winds, which arise from the vast stretches of the Indian Ocean, Arabian Sea and Bay of Bengal. They compel them to almost completely shed their water on the wind facing, India side of mountain ranges. The result is torrential rains and growth of dense forests on valley slopes on the Indian side but the northern Tibetan side falling in the rain shadow area is devoid of as much rainfall.

Transformation of Rocky Planet into Habitable Earth and Formation of Alluvial Plains

The transformation of rocky, inhabitable planet into a habitable one is a long story in which three primary forces of nature – the Sun, Water and Wind – allied together in their onslaughts on rocky Earth to carve out and form its physical features that make it host a variety of organic life, from microbes in the soil, variety of vegetation and human life, besides the habitats of water. The visible manifestations of their allied onslaughts are the following:

- Weathering is a process in which the rocky earth material is systematically broken into smaller and smaller pieces which cause reduction of mountains and formation of valleys.
- Erosion: This caused immediately imperceptible removal of the Earth material from the parent body. The process of erosion not only degraded the surface but the products of erosion facilitated further erosion which carved out the river the valleys, the riverine plains and determined the path of the river flow in the direction of the natural gradient of the strata. They carry the eroded material as sediment with their flowing waters.
- Fluvial action transports the eroded material and sediments under the influence of gravity, only to be further eroded and redeposited time and again until the residual material is deposited in the ocean.

Rain and river waters flow down the valley slopes in the Himadri and the Himalayan ranges southwards. The velocity of flow depends on the gradient and roughness of the bed. A steeper slope causes higher flow velocity but a rougher channel decreases it. Mountain streams begin with turbulent velocity that rolls down large pieces of rocks and detritus with itself to join with other streams to form a large river when *the flow mellows down as the flow flattens.* Whilst turbulent flow of water itself is capable of causing heavy erosion of the bed rocks of the river/stream, the rolling rock pieces accentuate the processes of erosion by scouring the bed rocks as well as through repeated strikes with each other.

There are two types of sediment materials in the river flow: first, those that are heavier and sink to the bottom of the river and roll down with the flow (called the bed flow); second, the sediment suspended or dissolved in the river water. They are transported by the water over long distances.

The Himalayan rivers carry a large load of sediments of varying sizes. According to the basic laws of hydrodynamics, their capacity to carry or push sediments depends on the velocity of water. Turbulent flow can push or roll down large pieces of rocks and boulders. They are gradually grounded during onward flow into shingle, cobble/pebble, gravel, sand and silt. The size-wise classification of sediments is given in Table 1.

Table 1

Classification of sediments according to size

Name	*Dimensions mm*	*Name*	*Dimensions mm*
Boulder	>256	Very Coarse Sand	02–01
Cobble	256–264	Coarse Sand	01–0.5
Very Coarse Gravel	064–032	Medium Coarse Sand	0.5–0.25
Coarse Gravel	036–016	Fine Sand	250–125 *
Medium Gravel	016–008	Very Fine Sand	125–62.5 *
Fine Gravel	008–004	Silt	62.5–3.9 *
Very Fine Gravel	004–002	Clay	<3.9 *

Note: colloidal < one micron. * unit is micron. 1 micron = 0.001 mm.

Thus, rainfall on mountains erodes and weathers their rocks, breaks them down to boulders, and gravel and other particulate, including alluvium, and soluble matter. This is a global phenomenon. It is estimated that globally rivers transport 17 billion tonnes of matter to the sea every year, 80 per cent particulate and 20 per cent soluble.

Formation of the Bhabhar and Terai Belts

As the streams emerge out of the steep Himalayan ranges into the plains, their flow velocity abruptly slows down. The streams are obliged to drop their heavier and bulky load, consisting of boulders, cobble/pebble, and gravel, over a narrow 10 km wide zone, called the Bhabhar belt. The accumulation of sediments below the Shivaliks of the Himalayan ranges gave rise to the formation of the Bhabhar belt. Since it consists of thus abandoned large pieces of stones, it is very porous and is, therefore, able to accommodate or quickly absorb large quantities of incoming water. The Shivaliks are often called the Himalayan foothills.

As the rivers emerge out of the Bhabhar belt, their flow is just fast enough to carry forward coarse sediments and silt. The gradient of its slope in the southerly direction is 55–75 mm/km. Just as the rivers enter the flat plains, the flow further slows down markedly as the gradient is reduced to about 6cm/km and a large amount of fine sediment and silt is deposited. The discarded sediments obstruct the flow to an extent that a huge marshy wetland is created, spread almost parallel along the Shivaliks. This wetland, often called the Terai, is another important habitat of fresh water. It is 26–32 km wide belt to the south of the Bhabhar Belt, spread between Yamuna river in the west and spreading to the east, running almost parallel to the Shivaliks.

The dense forest cover helped to maintain the physical integrity of the Terai land for millenniums of years. The fine network of roots of trees tightly held or locked the soil within their network; thus shielding the soil or the earth material from the erosive action of the running water. On emerging from The Terais, the rivers carry only silt and other fine suspended sediments. Technically, silt particles are so defined that they can remain suspended in water flowing at velocity as slow as 10 cm/hour.

The southern interface of the Terai Belt is with the alluvial plains of the Ganga basin which slopes very gently towards the east. Water is not able to exit the Terai region as expeditiously as it comes. This fact also facilitates the swampiness of the Terai region. In order to facilitate a rapid exit of their waters, some rivers split into a number of sub-streams which do not have a defined path; but emerging from this situation, they meet again to reconstitute the river. This pattern of changing flow path is shown in Figure 5 with reference to river Kosi which enters India through Nepal to unite with Ganga or Ganges.

Geological studies show that the sub-streams of Kosi have changed their paths 21 times since 1736 CE and Kosi has shifted its path about 113 km to the west.

Figure 5

Changing flow paths of the river Kosi over the years

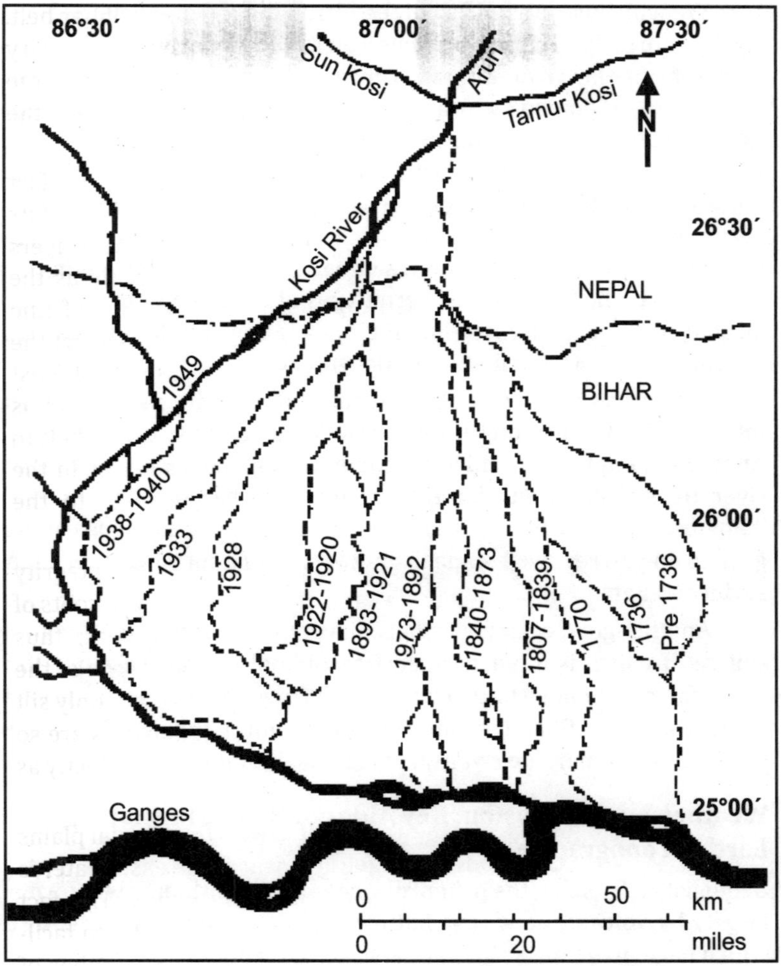

Bhabhar and Terai Belts: The Transit Stores for Redistribution of Himalayan Waters

As the porosity of the Bhabhar belt is very high, the water of the rivers is almost instantly stored in its hollow space for subsequent re-distribution downwards amongst other rivers. Water in the bowels of the Bhabhar hollowness builds up hydrostatic pressure which forces the water to come out as springs at geological faults or cracks or any other weakness in the down stratum. There are many such springs in northern regions adjacent to the Terais.

We will illustrate the fact of the Bhabhar and Terai belts acting as transit stores for Himalayan waters with a few examples. The first comes from the river Gomati. Gomati apparently originates from Gomat Taal, 32 km away to the east of the small town, Pilibhit, in the Terai region of Uttar Pradesh. In its first 35 km journey, the river appears and disappears frequently; it is only after flowing another 65 km and collecting the waters of its own tributaries, that Gomati becomes a well defined river. These specific features of the river have earned for it to be called 'the transcendental' river; thus giving it a divine aura. The 'ashrams' or hermitages of many sages of yore were located on its banks. After flowing for another 1,200 km, the river becomes yet another tributary of Ganga.

Similarly, the river Ghaghra, which enters India via Nepal, splits into two in Nepal which unite in India to reconstitute the river. Another example is provided by a tributary stream, Bata, of Yamuna river; the stream is considered to originate from a small stream called 'Jalmoosa ka Khala' which after flowing for some distance goes underground only to reappear some distance away as a spring to flow on the surface as Bata.

Another good example is that of river Choti Gandak, the apparent source of which is a place called Dhesophool in Maharajganj (Uttar Pradesh) but geologists believe that all its water comes from what went inside the Bhabhar and Terai belts. After flowing 250 km in serpentine manner, the river joins the Ghaghra in Bihar.

Weathering and Erosion: Key Aides of Nature in Shaping Earth's Topography

Geologists tell us that the primordial base of all continents was formed largely by volcanic activities; but most of the topographical features which have made the earth habitable were crafted by Nature through the processes of weathering and erosion of the rocks. These were caused by moving ice (glaciers and ice sheets), highly turbulent flow of water and wind. Erosion is the generic name for the acts which wear away or disintegrate the earth's strata and its mountain rocks. These acts occur through three kinds of weathering processes, namely mechanical, chemical and organic.

Mechanical weathering physically breaks up or causes fragmentation the solid rocks into pieces or chips. No rock is strong enough to withstand the forces of weathering and erosion. The chips of rocks are carried away by the flowing water as sediments.

Chemical weathering decomposes the rocks and minerals in the rocks; in the chemical processes, the structure of the rocks is chemically altered and thus destroyed; one simple example is water dissolving away soluble minerals from limestone rocks, the large spaces occupied by water soluble compounds or minerals become caverns in the limestone stratum. They ordinarily become storehouses for underground water or when exposed to the surface ponds and lakes (they are further discussed later in the book). Organic weathering happens when plants break up rocks with their growing roots or when the organic acids formed on putrefaction of vegetative matter dissolve the rock material. Together these forces have sculpted the peaks, deep gorges and valleys of the Himalaya mountains. They have effectively shaped the topography of the surface of the Earth.

It is important to note that Nature itself provides, in situ, the forces that cause physical fragmentation or chemical disintegration of rocks.

Erosion by Ice

We are also told by geologists that geological ages back, the climate of the Earth was not as it is now. Earth has witnessed 'ice ages' during which parts of the northern hemisphere were covered with ice and vast glaciers roamed on the surface and on frigid mountain tops, slowly moving downhill across the land. Moving glaciers with embedded rock pieces in their ice heavily scour or erode their base; the sediments which move with the glaciers could be sharp edged rock pieces varying in size from huge boulders to tiny grains of sand. These rub against the ground below, eroding both the ground and the embedded rock piece. The glacier sediments are known as 'morains'. The erosive action of glaciers over ages has resulted in the formation of gorges in Himalayas and other mountains, fjords along the coast of Scandinavia and several lakes or has paved riverine valleys and plains.

Erosion by Water

Among ice, water and wind, running water is the most effective agent of erosion. It not only carries heavy sediments but can wear away the banks of the rivers thereby widening the valleys. The continuous battering of mighty sea waves against the shoreline erodes the lands and helps the sea to encroach upon the land by taking away the sand from the beaches. When a river slows down at its mouth, the fine sediment that it has transported over long distances is deposited in a new location, i.e. at the interface with the sea. As the sediment builds up, it creates fertile land, called delta. River deltas are made up almost entirely of silty alluvium (sediment). In this manner, the land encroaches upon the sea. At other places where the sea currents are strong, or the velocity of river flow is high, the currents carry away the silt away far into the sea. The deposition of silt in the form of delta does not occur. Streams from tiny creeks to huge rivers carry tonnes of eroded soil and rocks every year. Beach erosion by the sea is a serious problem at many places where the sea shore is made of smaller particles, like sand which are easily washed away by powerful waves and currents or of softer rocks from which only tiny bits are removed at a time by the erosive processes.

Erosion by Wind

Wind by itself may not cause any significant erosion of hard rocks. However, it, almost always, carries some dust or sand particles. When the wind blown sand particles strike against a rock or a man-made structure, it wears away the rock so slowly that its relief pattern or roughness is smoothened. It is the weakest of the three agents of erosion. In environments where the atmosphere contains carbon dioxide gas, the erosive action, particularly on sandstones and marble, is aggravated or accentuated because of the formation of carbonic acid when carbon dioxide dissolves in the moisture of the air.

Mechanical Processes of Fragmentation

There are several ways through which small chips from rocky body can be removed mechanically. It is well known that when two brittle bodies, moving randomly in different directions, strike together then they may shatter into small pieces. In the same manner, rain drops (including waterfalls) and hail stones have been incessantly falling on mountain rocks from the great heights of the clouds; the forces created by the impact dislodge minute particles from the rocky substance. Howsoever imperceptible the loss of the rock body through erosive chipping may be in the short run, the loss is certainly perceptible over a period of time. Flowing water immediately carries away the finely chopped debris and deposits it elsewhere in accordance with the laws of hydrodynamics.

Similarly, mechanical erosion also occurs when boulders, pebbles and other rock pieces strike against one another during their flow with turbulent waters of streams/rivers; in the process, the rough surfaces of rock pieces are smoothened by their repeated strikes against each other during their downward movement with the turbulent stream. It may be noted that mechanical erosion does not bring about any chemical change in the eroded particulates. Nevertheless, the surface area increases manifolds.

Mechanical erosion facilitates chemical attack. Since chemical attack is initiated at the surface only, fine particles facilitate chemical attack on them when environmental conditions are otherwise favourable. It is through mild chemical attack in the environs of flowing river that oxygen dissolved in river water converts the inert rock particles (sand) into particulates of soil/alluvium through complex chemical reactions (described later). The important point to note is that, other environmental conditions remaining the same, the rock body cannot change into soil though it has the same chemical composition as its own eroded particles. The fineness of the rock particle is, thus, central to its conversion into particles into soil. We shall know in a later chapter that the water holding capacity of soil is greater than that of rock particles or sand.

Other Factors Resulting in Mechanical Erosion

Other important factors promoting disintegration of rocks are:

1. Repeated freezing and melting of water filled in the cracks in the rock body on account of frequent atmospheric temperature fluctuations around the freezing temperature, i.e. 0°C. Since there is about 9 per cent volume expansion on freezing, tensile stresses of high magnitude are set up at the crack tip, making the crack grow and finally making the rock body to split. Such temperature fluctuations are not uncommon in upper Himalayan ranges. This is called the frost shattering. Every time the physical process occurs, the cracks are opened a little more wide. Over time, large pieces of rocks can split off into smaller rocks.
2. Forest fires raise the surface temperatures of the rocky stratum causing it to expand relatively to the deeper layers in the stratum. Repeated occurrence of this phenomenon causes rocks to crack and disintegrate. Similar situation is created by the solar heating of top surface cooling down of the rock body in day time and its cooling down in the night.
3. A rock usually has more than one mineral in its body. Each mineral has its own characteristic coefficient of thermal

expansion. Temperature induced expansion (or contraction) sets up situations of differential expansion distribution on micro scale within the mass of the rock body. This closely mimics the situation explained in the point above.

4. Germination of trees and growth of their roots within the crack is a frequently observed phenomenon; the thickened roots act as a wedge to split open the rock. This phenomenon can also be observed in neglected buildings made of bricks and mortar.

Chemical Processes of Weathering

Water is ordinarily described as chemically inert or neutral but in strict scientific terms, this statement is a myth. Its electronic structure is such that under certain situations it behaves like a weak acid. As the weak acidic rain drops dissolve carbon dioxide present in the atmosphere, their acidity is accentuated; rain drops act as weak carbonic acid. When such acidified water flows over carbonate rocks, namely limestone, the rocks are subjected to chemical erosion. Further, if colonies of microbes are flourishing within the crack, the carbon dioxide from their respiration process may also strengthen the rain water acidity and thus increase the rate of decay of carbonate rocks. The mysterious ways through which Nature works are slowly being unmasked by science.

Other weak organic acids produced during decaying of organic matter exercise similar effects. Rock mineral particles, hydrated in presence of water, together with the decayed organic matter make the alluvial soil, usually in the ratio 90:10. The formation of soil from disintegrated particles of rocks is in itself a matter of deep studies, our intention in describing this is to draw attention to the fact that in the process of making the alluvial plains (which are also storehouse for underground water), mechanical and chemical processes of weathering have played a central role.

Erosion is a natural process which occurs slowly and gradually over time (millions, thousands or hundreds of years) but which effectively changes the topography and landscape by wearing down mountains, creating valleys or destroying them, making rivers appear or disappear. As explained above, environmental factors break rocks into smaller pieces and loosen them from the parent rock/structure, only to be carried away by water and wind to be deposited elsewhere.

Effect of Anthropogenic Activities on Erosion

Experts in ocean sciences are of the opinion that the increase in temperature around the world on account of global warming is speeding up erosion in several ways. Prominently through:

- More frequent and more severe storms - hurricanes, typhoons, tornadoes - and tsunamis threaten to erode miles or kilometres of coastlines and coastal habitats.
- The enhanced environmental temperatures are causing faster melting of polar ice caps. It is causing sea levels to rise. It is estimated that a rise in sea levels of 8–10 centimetres in the sea level in the eastern United States will cause enough erosion to threaten buildings, roads, tunnels and sewer systems.
- Rapid melting of glaciers has threatened the existence of many lakes in the upstream of morain dams. In fact a few cases of bursting of morain dams in the Himalayas have caused disastrous floods in the downward flow direction of rivers.
- Melting of the permafrost in the Himalayas which has resulted in frequent land slides and flash floods which cause severe erosion on mountain slopes and increase the sediment load of the streams and rivers.
- Cutting down of trees and plants. They hold soil in place and prevent the soil from being washed away more easily. Landslides and flash floods become more common as rain water also rushes over exposed soil rather than soaking into it, causing flooding.

Water Habitats in and those Created through Himalayas

The running water of the streams also brings down the rotting organic matter from the dense forests. As the Himalayan ranges continued to rise from the sedimentary rocks, the organic matter enriched soil too continued to be deposited on the bed of the contracting Tethys Sea and eventually made up the largest and agriculturally the most productive alluvium plain in the world. We hasten to point out that this alluvium plain is also the most important source of underground water in India. Thus, the Himalayas have created four best known habitats of fresh water, namely:

- Permanent bed of ice, glaciers and permafrost
- Highly fragmented sedimentary rocks which provide residence space to water (further discussion in another chapter)
- The Bhabhar and Terai belts
- The alluvium plains

Visible Evidence for the Natural Occurrence of Weathering Phenomena

The phenomena of weathering and fluvial action occurring over millennium years have gifted us the best and most easily accessible underground habitats for fresh water all over the world. Weathering involves all primary elements of Nature – wind, water, sun and earthy matter. These are universally present and operative all the time – some take a short time to show themselves having occurred, others may take a longer time. India has a number of archaeological monuments of great architectural beauty standing exposed to the elements of nature for centuries. Careful scrutiny of their carvings would show that many of their intricately carved features, particularly those standing in relief facing the windward side have eroded away. This has happened to the carvings of the Sun Temples at Konark in Odisha or in Gujarat, or to those of Khajuraho in Madhya Pradesh or to Qutub Minar in Delhi.

Effects of chemical weathering manifest in much shorter duration of time. We are aware of the presence of small amounts of carbon dioxide and of sulphur dioxide in the atmosphere either naturally or on account of industrial activities. Along with the moisture in the air/atmosphere, they form weak acids which aggravate the erosion process by adding a chemical component to the mechanical process. Chemico-mechanical weathering may also be seen occurring on neglected buildings. Moss begins to grow silently at damp walls. On removing the moss layer, one can see effects of chemical weathering on the walls.

Classification of Indian Soils: Their Water Holding Capacities and Distribution

We have seen that Nature created fresh water from the vast oceans of saline water through rains. Habitats or storage spaces for fresh water were first created through very violent and catastrophic geological events, through which the high mountains and rivers/lakes were formed. Subsequently, the slowly occurring twin phenomena of erosion and weathering took over. Thus, the north Indian alluvial plains formed very slowly to become one of the largest storehouses for fresh water.

The first contact of rain drops on touching the ground is with the surface. The options available to it are: (a) to immediately evaporate, (b) to run along the slope of the surface to join rivulets/rivers with the intention of returning to the sea, or (c) to infiltrate into the ground and hunt for underground resting places or habitats. In fact, it does all the three.

In so far as its tendency to infiltrate is concerned, the nature of the top surface stratum (immediately underneath the surface) plays a pivotal role in temporarily holding the rain water within its 'folds'; thus restricting the amount of 'run off' water. Thereafter, the geological nature of the underground strata determines the nature of its underground habitats. If the porosity and permeability of strata are poor (explained in chapter on underground water), much of rain water, other than that lost through evaporation, flows back to the sea. Whilst these issues will be discussed in later chapters, we are presenting an account of the nature of top soil here as it is the first acquaintance of the rain water with earth after it has fallen on the ground. One thing is certain that all the top soils of India are not of alluvial origin but have formed from the rocks of the area through other processes of weathering.

There can be a few methods of classifying the top soil. For a farmer, the soil is either fertile or infertile or even barren. Another method of classifying the soil takes into cognizance the nature of rocks from which it has formed and the manner through which the soil had formed. As with the formation of alluvial soils, the formation of soils on the strata of igneous and crystalline rocks also occurred through 'weathering'. Their classification on this basis is shown in Figure 6. Important classes of the soil are:

- Alluvial Soil
- Black Cotton Soil
- Red and Yellow Soil
- Lateritic Soil
- Arid and Desert Soil
- Saline Soil
- Mountain Soil
- Peaty Soil or Bogs

Alluvial Soils

Apart from the north Indian alluvium plains, alluvial soil is found in many parts of India along all river valleys, notably of Kaveri, and in their deltas; also along the coastal regions along the Western and Eastern Ghats but these river valley alluvial soil regions are not shown in Figure 6. Almost 40 per cent of the Indian land area is covered by

Figure 6

Distribution of different types of soils in India

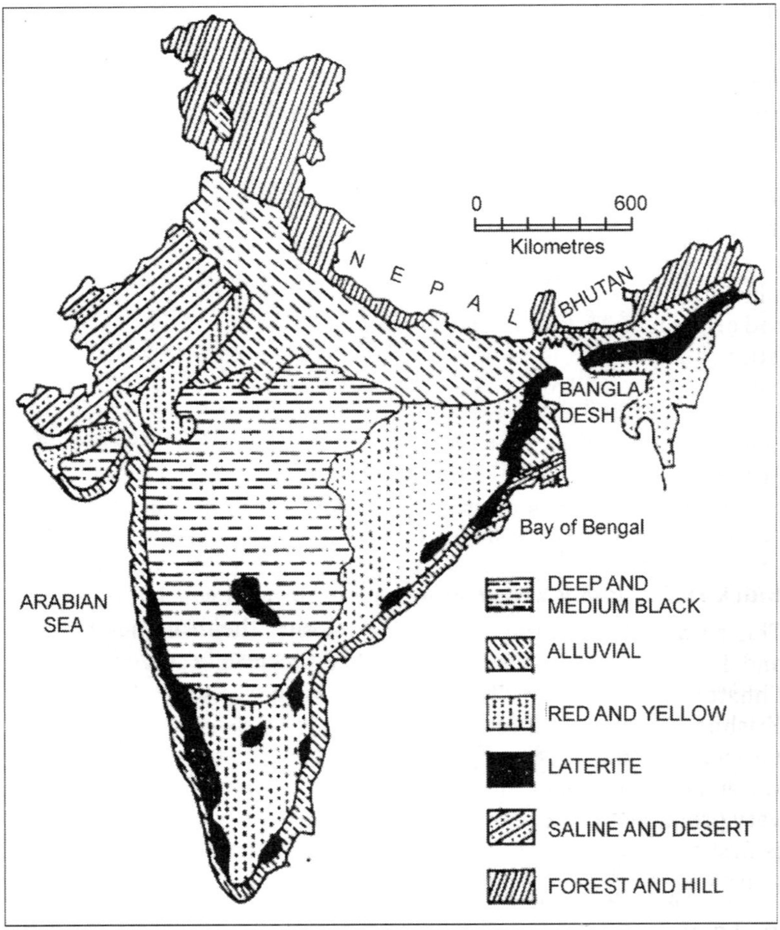

Source: Food and Agricultural Organisation of the United Nations, http://www.fao.org/docrep/003/V5930E/V5930E01.htm

alluvial soil. All alluvial plains are characterized by their great capacity to hold water and high permeability. These basic characteristics make them ideal hosts for ground water. Water is available in wells of shallow depths and they are easy to recharge with rain water.

There is, however, a basic difference in the manner through which the alluvium accumulated in its north and south Indian habitats. The north Indian rivers had the deeps of the erstwhile Tethys

Sea available to them to fill up with their alluvial sediments but the south Indian rivers had first to pave their way through the hard rocks of the Deccan Plateau. They had no natural place available to deposit their sediments. Further, they had much less sediment load to carry than their Himalayan counterparts. The Himalayan rocks were comparatively much more badly shattered than the rock strata of the Ghats (by the geological activities). Shattered pieces of rocks were readily available in Himalayan valleys to roll down with the torrential flow of water. Therefore, the north Indian plains have great depths but those of the south have shallow depths.

It may also be appreciated that the chemical nature of any alluvial soil, would carry the DNA of the parent rocks, from which it had originated. Thus its chemical nature and other physical characteristics could vary from region to region, i.e. the alluvial soil in Kaveri basin is different from that of Godavari or Mahanadi basins. Such variations would also exist within the great north Indian alluvial plains as its northern parts have formed from the sediments carried by the Himalayan rivers and the southern regions by the sediments of rivers originating from parts of the central Indian highlands, namely, Chambal, Son, Damodar, Subernrekha, etc.

Black Soil or Black Cotton Soil

This soil has been formed through the weathering of igneous rocks and is spread over many parts of Maharashtra, parts of Gujarat, Chhattisgarh and Telangana. It is also found in the valleys of Godavari, Krishna, Narmada and Tapti. It has shallow beds, about 6 m deep. Its permeability is poor; being porous, it has tremendous capacity to hold water within its mass. It swells up when holding water; when dry, very prominent cracks are formed. In the event of drought, its surface is first to show these cracks. It is, therefore, a hot favourite with the journalists to show the state of drought.

Red Soil

This soil has been formed through the weathering of crystalline and metamorphic rocks. It is porous and permeable. Therefore, its water holding capacity is poor. It is spread over Tamil Nadu, south Karnataka, north-east Andhra Pradesh and Odisha. Its red colour is suggestive of the presence of iron but the colour changes on hydration.

Lateritic Soil

This soil is found most prominently in Western Ghats, in several pockets in Deccan plateau and in isolated pockets in many other parts of India. Chemically, it has hydroxides of iron and aluminium with some clay. Its formation occurs through chemical weathering in presence of water at modestly high temperatures. Therefore this type of soil is found in regions within the tropics. It has tremendous capacity to hold water within itself. However, the content of humus and organics in its mass is poor because it provides ideal conditions of humidity and warmth for bacterial growth.

Arid, Desert and Saline Soils

Desert, sandy soil is found in western Rajasthan in areas where the rainfall is scanty. Sandy soils are permeable. Therefore, whatever little rains falls, it easily flows out in the direction of natural slope of the stratum. If another impermeable stratum lies immediately under the sandy bed, flood-like situations could easily develop. It is interesting to note that this character of soil and local geology was, perhaps, known to the local people who had designed systems to collect the rain water for societal use in earlier centuries.

Soils are rendered saline through the ingress of sea water. Necessarily, these types of soils occur in river deltas near their interface with the sea or through oozing out of saline underground water. Whilst the sea tries to claim more sweet soil to turn it saline, Nature sides with the so-called sweet soil to thwart the attempts of the sea to ingress further by planting vegetation which survives in saline water, particularly the mangroves. Uniquely, roots of new mangrove tree begin to grow whilst the fruit is still on the mother tree. This act helps to make the mangrove forests very dense, which effectively restrain the gushing sea water from running up the streams of fresh water.

In modern times, good soils are turning saline through excessive irrigation. In many areas of Punjab and Haryana, irrigation has become counter productive as it has promoted the soil to turn saline. Increasingly greater percolation of water in the ground raises the water table. With this, the salts from the deeper underground water move upwards and into the vadose zone. The situation is bad but the trend towards salinity can be reversed by scientific management of irrigation practices. This fact heightens the need for spreading scientific awareness about fresh water and its resources.

Western Ghats or the Sahyadri Mountain Ranges: Abode of Fresh Water

What the Himalayas are to north India as the divine water bank, so are the Sahyadris for the entire peninsular India. Their geographic spread always remains very close to the Arabian Sea all along the western coast. It begins from the Tapi river in the north and culminates with the very southern tip of India, Kanyakumari, covering a length of about 1,490 km with width of about 240 km and area of 1,29,037 sq. km in the states of Gujarat, Maharashtra, Goa, Karnataka, Kerala and Tamil Nadu and home to about 50 million people. They arise steeply from the Arabian Sea as if to protect the peninsular India from the onslaughts of the sea. The sea, however, extracts a heavy price from the Sahyadris for sheltering the land – its perpetual enemy – in the form of fresh water which the clouds have given them after extracting the same from the sea through evaporation. The Sahyadris do so through hundreds of minor streams running down their western slopes.

Climatically, this zone of hot tropical climate is blessed with good rainfall which, however, is confined to about four months from June to September. The magnitude of rainfall varies a great deal across its span; from as high as 7,000 mm in Kudremukh area to about 3,000–5,000 mm in other areas. Their geographical dispensation from north to south is such that it faces Monsoon Wind on its west and rainshadow areas on the east. Yet the rainfall in the rainshadow areas could be significantly higher than many other parts of India, at about 900–1,500 mm per annum.

The mountain ranges are not as high as the Himalayas but high enough to influence the distribution of monsoon rains in India. The overall natural slope of the Deccan plateau is from west to east. Many prominent rivers of South India – notably Godavari, Krishna, Kaveri, Tamraparini, Vaitarna, Netravati, Tungbhadra, and others – have their sources in the eastern or the rain shadow side of the Sahyadris or the Western Ghats, flow easterly but the much lower Eastern Ghats block their passage; the rivers have to meander over the plateau to locate the gaps in the Eastern Ghats so that they could discharge their waters into the Bay of Bengal. They carried water for most part of the year along their entire path, until recently when many reservoirs built in the river path began taking away their waters. Important question is: what is the source of their near perennial water supply?

The mystery is resolved by the following two considerations:

- In many areas of the Western Ghats or the Sahyadri ranges, the top stratum consists of lateritic soil which has enormous capacity to hold rain water as it comes, until it slowly penetrates into the lower strata. In this manner, the lateritic soil cover acts merely as a transit store.

- Their strata slope towards the east. Thus a good part of the rain water falling on the Sahyadri surface eventually flows into the east flowing rivers.
- Elsewhere, the mountain valleys are covered with dense evergreen forests and grass. They variously promote or facilitate the infiltration of water into the ground. Their root network wrapping around aggregates of soil particles forms a kind of sponge to hold water for a longer time and, thus, to provide longer time for the process of infiltration.

Types of Vegetation Specially Useful in Promoting Build-Up of Underground Water Resources

The slopes in the valleys are fully covered by trees which are specially adapted to the local environmental conditions. The following types of forest cover are especially helpful in this regard:

1. Evergreen forests or wet montane evergreen forests occur at elevations of 1,200–2,200 m in the southern regions of the Ghats in the undulating valleys. At elevations higher than 1,600 m grasslands of perennial Shola grass reinforce the water holding capacity of the soil. The forests are very dense. Dense fog occurs very frequently in the region.
2. Semi evergreen forests found especially in lateritic soils of Goa, Maharashtra and Karnataka.
3. Moist deciduous forests are found in the southern parts of the ranges and in Nilgiris at elevations of 1,000 m with rainfall of 2,500–3,500 mm and cover the largest area of the forest cover Because of dense forests, heavy rains, fog and high moisture content of the environment, the trees are covered with moss, algi, ferns and other woody climbers, on the one hand and on the other, they help to create bogs on the ground; bogs are a kind of swamp made by rotting vegetation, including trunks and roots of the trees with water. Bogs can be seen as a natural sponge on the surface of the Earth. It enables water to stay on the surface for longer time and so infiltration of water in the ground occurs more effectively.

Important point to note is that the forest cover on the Western Ghats is as important as the mountains themselves for the purpose of their continuing to remain as the 'water bucket' of peninsular India. Both of them are under increasingly greater threats of destruction through human activities – the mountains through mining for iron ores and forest cover through deforestation for human habitation and other related activities. Ancient Indians had used the traditional Indian belief in god to protect the forests. Many of the forests were declared as 'sacred groves' so that they could not be cut and only dead wood could be carried away as firewood. In the same vein, some trees were specifically associated with different manifestations of the Supreme – Banyan tree with Lord Vishnu, Bel and Rudraksh with Lord Shiva, etc.

The Western Ghats are a national treasure of water resources for almost the entire southern India. Their virginity and sanctity must be preserved in order to provide water security to our population in parts of Maharashtra, Karnataka, Telangana, Andhra Pradesh and Tamil Nadu. We must all join together to protect our natural heritage in the form of topography and forests as they are central to the preservation of our primary resources of water.

References

Anticline, https://en.wikipedia.org/wiki/Anticline.

Classification of Sediments, Sediment, Wikipedia https://en.wikipedia.org/wiki/Sediment.

Continent-continent Convergent Plate Boundaries, http://www.ck12.org/earth-science/science/Continent-Continent-Convergent-Plate- Boundaries/lesson/Continent-ContinentConvergent-Plate-Boundaries-HS-ES / 7.

Ecology of the Himalaya, Wikipedia, https://en.wikipedia.org/wiki/Ecology_of_the_Himalaya.

Erosion, National Geographic, http://nationalgeographic.org/encyclopedia/erosion/

Food and Agricultural Organisation of the United Nations, http://www.fao.org/docrep/003/V5930E/V5930E01.htm

Geography of Himalayas, Terai, Bhabar, Shivalik, Lesser and Greater ... http://www.gktoday.in/blog/geography-of-himalayas-terai-bhabar-shivalik-lesser-and-greater-himalayas/.

Himalayas, When two continents Collide: http://pubs.usgs.gov/gip/dynamic/himalaya.html.

Indian Plate, Wikipedia: https://en.wikipedia.org/wiki/Indian_Plate

Interactives, Dynamic Earth. Plates & Boundaries – Annenberg Learner http://www.learner.org/interactives/.../plate.html

Joshii, V.K., R. evnge of Kosi, http://www.boloji.com/index.cfm?md=Contentsd=Articles&ArticleID=4701 (since withdrawn) alternate Dinesh Kumar Misra, The Legends of Kosi, march 2009, http://old.himalmag.com/component/content/article/468-the-legends-of-kosi -html.

Kumar, Gopendra, 2005, *Geology of Uttar Pradesh & Uttaranchal*, Geological Society of India. Bangalore.

Landforms of fluvial erosion and deposition – Ace Geography http://www.acegeography.com/landforms-of-fluvial-erosion-and-deposition.html.

Metamorphic Rocks, Tulane University, www.tulane.edu/~sanelson/... /metamorphic.htm.

Mountains and Deformation of Rock, Tulane University, www.tulane.edu/~sanelson/eens1110/deform.htm.

Plate Tectonics, Tulane University, www.tulane.edu/~sanelson/eens1110/pltect.htm.

Plate Tectonics, Wikipedia: https://en.wikipedia.org/wiki/Plate_tectonics

Social Science (Std 10) Govt. of Tamil Nadu) https://archive.org/details/std10socsciem2011tnbooks.

Sugunun, V.V., Reservoir Fisheries of India, Food & Agricultural of the UN 1995. http://www.fao.org/docrep/003/v5930e/v5930e01.htm.

Tectonic Plates Map and Information Page, World Atlas www.worldatlas.com/aatlas/infopage/tectonic.htm

Teraii, Wikipedia https://en.wikipedia.org/wiki/Terai.

The Fluvial Landforms and Cycle of Erosion, Geography Notes www.geographynotes.com/.../the-fluvial...of.../757.

Weathering, National Geographic Society, http://nationalgeographic.org/encyclopedia/weathering/.

Chapter 4

Global Distribution of Water

Water is widely distributed on earth. The Earth appears with blue hallow when observed from far up in the space. It is now being glorified as the only Blue Planet in the solar system. Water is distributed on earth in all the three states of matter, i.e. solid ice, liquid water and vapour in air. Water plays a central role in defining the climate of any place. In association with solar heat and wind, it has carved the geomorphological features of earth. It has facilitated the formation of fertile alluvial plains.

Most of it, almost 97 per cent of the total quantity of water available on earth, is saline. It is distributed in large, mutually interconnected oceanic bodies which are known by their different names – Pacific Ocean, Atlantic Ocean, Indian Ocean, Antarctic Ocean (the South Sea) and the Arctic Ocean. In addition, there are bodies of saline water within the land mass, such as the Caspian Sea, Dead Sea and the Black Sea or the Sambhar Lake in India. A fairly good amount of ground water is also saline. It is important to know that only liquid water can be saline but its gaseous states, vapour and steam, are always of pure water. Similarly in its naturally occurring solid state, i.e. ice, it is always pure water.

Salt free, or the so-called fresh water constitutes less than 3 per cent of the total quantity on earth but it is largely present as the polar ice caps on the continent of Antarctica and Greenland. Fresh water, available to mankind for direct use, may be less than 1 per cent of the total and is made available through rains or in rivers/lakes and underground resources.

Man's first familiarity with the vapour state was through the condensation of the vapour into drops of rains. He soon gained intimacy with steam – another manifestation of gaseous state. His understanding of the properties of steam and their engineering potential heralded the industrial revolution. Apart from being used for drinking and other domestic applications, fresh water is increasingly being used for irrigation, municipal or in industrial applications, including thermal power generation. The demand for fresh water is continuously increasing but its natural resources are limited to quantity that was available when the earth became habitable. There is no doubt that water was plentifully available till in the recent past but it is not so today. Its decreased availability is not because the quantity available from its natural habitats has decreased through some cataclysmic geological event/s. Indeed, its renewable total quantity available today is the same as it was thousands of years ago. The scarcity in its availability is felt because of tremendous increase in global population and increased demands of water on account of its use in industry and irrigation. There has been quantum jump in area of farmlands to provide food security to billions of people, many of whom are still in the hungry zone of living below the minimum subsistence levels of food consumption.

Distribution of Water among its Natural Habitats

Water is that extraordinarily unique creation of Nature that it is indestructible; though it easily changes the manner of its manifestation. The same water is in human body at one time; at another in the body of an animal or insect or a grain of wheat/millet/corn and at still another time, it may be providing motive power to drive wheels of industry or irrigating farm fields. Water merely changes its form. Despite continued use over thousands of years, its available quantity has remained the same, courtesy the global hydrological water cycle.

Table 1

Distribution of globally available water in its different habitats

Habitats	*Volume cu. km*	*Volume cu. km*	*% Total Water*
All Oceans	**1,33,80,00,000**		**96.5**
Pacific		66,98,80,000	48.30
Atlantic		3,10,410.000	22.40
Indian		2,64,000.000	19.0
South		7,10,00,000	5.18
Arctic		1,87,50,000	1.35

Cont'd...

...Cont'd

Ice, Snow and Permafrost	**2,43,64,000**		**1.76**
Icecaps Glaciers		2,40,64,000	1.74
Antarctica Ice			1.62
Greenland Ice		22,34,000	0.17
Glaciers		2,40,000	0.017
Ice, Snow on Land		3,00,000	0.022
Underground Water	**2,40,64,000**		**1.74**
Saline		1,28,70,000	0.93
Fresh (salt free)		1,05,30,000	0.76
Ground Moisture	**16,500**		**0.0012**
All Lakes on Land Surface	**1,76,000**		**0.013**
All Saline Lakes, incl. Caspian		85,400	
All Fresh Water Lakes		91,000	0.0066
Swamps	11,470		0.00083
Rivers	2,120		0.00015
Biospheric Water	1,120		0.00008
Atmosphere	12,900		0.00093

Source: Igor Shiklomanov's chapter 'World Fresh Water Resources' in Peter H. Gleick (editor), 1993, *Water in Crisis: A Guide to the World's Fresh Water Resources* (Oxford University Press, New Yor).

The total quantity of available water always remains distributed among its different natural habitats in almost the same proportions as indicated in Table 1. But small amounts from each of the habitats keep changing their locations, almost on barter basis, i.e. if the sea loses some amount through evaporation, it gets back almost the same quantity from the rivers; or when a small amount of the ice on the Antarctica continent melts, an equivalent amount comes back as snow. In other words, there is complete harmony amongst its different habitats. (But global warming has begun to change this harmony). Its systematic renewability distinguishes water from other natural resources, which once used are destroyed forever; coal, petrol or diesel oil once combusted cannot be recreated. Or the iron ore from which iron has been extracted cannot be recreated. In this sense, water is immortal.

The figures indicated in Table 1 are estimates arrived at by specialists and are not subject to verification by actual measurements. Thus, estimates by different specialists may not tally quantitatively with each other but there is remarkable agreement in their relative distribution patterns.

Table 1 shows that fresh water is about 3 per cent of the total quantity of water available on the planet Earth. This amount is distributed among its different habitats as shown in Table 2(a).

Table 2(a)

Distribution of fresh water in its different types of habitats

Habitat	*% of total fresh water*	*Habitat*	*% of total fresh water*
Ice Caps	68.7%	Permafrost	0.8%
Underground water	30.1%	Surface Water	0.4%

Total surface fresh water further distributed among its different habitats as shown in Tables 2(b).

Table 2(b)

Habitat	*% of total fresh water*	*Habitat*	*% of total fresh water*
Lakes	67.4%	In Marshy Lands	8.5%
Soil Moisture	12.2%	Biosphere	0.6%
Rivers	1.2%	Miscellaneous	0.6%
Moisture in Air	9.5%		

Thus mankind has only a minuscule quantity of the total water available to manage its life and all its activities, though in recent times mankind has begun supplementing the quantity of fresh water available from natural resources through the technology of desalination of sea water.

Residence Time of Water in Different Habitats

Thus, it is clear that different habitats are not watertight compartments. They are, indeed, autonomous components in the Commonwealth of Water. Parts of it keep exchanging their habitat from one to another through the hydrological cycle. If an individual molecule could be marked for identification, it may be found in sea at one point of time, as snow on Himalayas at another or in human body at yet another time or deep underground. Experts have determined the average residence time of any water molecule in its various habitats. These are shown in Table 3.

Table 3

Average residence time of water in different habitats

Habitat	*Residence*	*Habitat*	*Residence*
Oceans	4,000 years	Underground Water	1–1,00,000 years
Lakes	10 years	Soil Moisture	4–50 years
Marshes	1–10 years	Ice Caps	10–1,000 years
Rivers	14 days	Atmosphere	10 days
Human Beings	7 days		

The data gives us several important hints so that we treat it respectfully and accord due sanctity to its habitats. The atmosphere is ever so keen to provide us a continuing supply of fresh water that it just takes a maximum of about 10 days to bring clouds formed from new evaporation from the sea to the land for raining. In the process, it leaves behind in the sea, all the pollutants which rivers have discharged into the sea. As a consequence thereof, the nature of sea water has already begun to change. The changes have begun to affect marine ecology. (These will be discussed ahead in the book). Once the pollutants enter the sea, they have no other place to go, evaporation only takes away the pure water component of sea water, leaving behind its salts and pollutant chemicals in the sea. The polluted water keeps inflowing into the seas through the rivers.

Similarly, some of the pollutants enter ground water through the process of infiltration. The rate of infiltration could be rapid at the top stratum of alluvial soil but it slows down deeper down. Our ancestors did not pollute the infiltrating ground water so that we are drawing water of good quality from both shallow and deep bore wells. Already the pollution level in waters of shallow wells is high enough in some parts of China and India that the water is unsuitable for human consumption. During the course of next few decades, polluted water of our making would reach deeper levels. It is a matter for serious introspection for us to think about the kind of underground water legacy we are leaving behind for our progeny.

The residence times in aquifers are shown to be up to 1,00,000 years. As we shall be seeing later, such aquifers exist in deep non-alluvial strata of the earth. In some parts of the world, water which had accumulated in them more than 20,000 years ago has been tapped to exhaustion in the last two or three decades. While many governments have now launched programmes for ground water recharging, there is no possibility of recharging these aquifers which experts often call *fossil aquifers*.

Oceans as Habitats of Water

Water in the oceans has been there ever since the planet cooled down sufficiently to be able to host it in its liquid form. Sea water is always in motion in various ways; some are obvious and others, not so. We are familiar with its movement in the form of visible surface waves and the periodic rises and falls in its levels, called tides. Tides are caused (as a first approximation only) by the gravitational pull exercised by the

Moon on large water bodies on the Earth. As a consequence, the oceans bulge out in the sublunar and antipodal directions of the Moon by simultaneously thinning out in out in their perpendicular directions.

In addition, strong ocean currents flow in oceans, much in the manner of rivers flowing on the surface of the earth. They carry large quantities of sea water from one region of the world to another. Their flow influences the climates of the regions. One prominent among such currents is the Gulf Stream which carries warm waters from the Gulf of Mexico to sweep past the coasts of North-West Europe. The solar energy, stored in the form of thermal energy in its warm waters, makes the climate of the region much milder than elsewhere of equivalent latitudes. Gulf Stream moves at speeds of 60–97 km per day carrying warm waters of tropical seas to West Europe and keeps it warm. It is estimated that the Gulf Stream moves 100 times the water as all the rivers of the world put together do.

Another of currents is the El Nino which rises off the coast of Peru in South America and moves towards the Pacific Ocean. It influences the monsoon rains in India and other countries of South and South-East Asia. Indeed, the Indian economy is so much dependent on the El Nino dependent rains that the El Nino, albeit Monsoon, could be referred as the virtual finance minister of India. The movement of the El Nino Current is subject of deep international studies with mutual cooperation. The largest ocean current, however, is the Kuroshio Current, off the coast of Japan, which can move with an ocean bed 1,000 m deep at a velocity varying between 40–120 km/day.

In addition massive currents could be set up in oceans on account of differences in the densities of sea waters caused by differences in their salinity and temperatures. Knowledge of basic science tells us that higher the temperature, lower the density but higher the salinity, greater is the density. As the Gulf Stream carries the warm waters of the Atlantic Ocean to North-Western European seas, denser and the super salty water from the Mediterranean Sea moves in as a replacement. Sea water movements inspired by differences in density and/or temperatures are known as thermohaline currents and a matter of intense studies by ocean scientists. As a result of these movements, erosion of under sea outcrops of rocky strata has been observed to occur. Ocean scientists have ample evidence to confirm its occurrence.

Many climatologists regard the oceans as a vast and expansive store house of solar energy and regulator of global climate. Indeed, considerable amount of sea water continuously enters the atmosphere through the solar energy inspired process of evaporation and transpiration over the millennium years.

What has contributed to its immortality is the fact that the land mass on the earth which receives the evaporated water in the form of rain, returns it to the sea in equal measure. The water, thus, entering the sea through the rivers is called 'runoff' water. In addition to the surface runoff waters, the seas also receive water from underground water flows.

The quantity of runoff water entering the sea from different continents is shown in Table 4. The data shows that run off vary greatly within continents. The largest amount of runoff water is coming from the largest and most populated continent of Asia and comparatively very little from North Africa, Middle East and Australia. Of course, the extent of 'runoff' provides an indication of the regional rainfall; higher the rainfall in the river basin, greater is the runoff.

Table 4

Runoff water entering sea from different continents

Continent	*Run Off Water, cu. km/year*	*% of Global Run Off Quantity*
North America	7,800	17.9
South America	12,000	27.6
Europe	2,900	6.7
North Africa and Middle East	140	0.3
Sub Sahara Africa	4,000	9.2
Asia excluding Middle East	13,000	30.6
Australia	440	1.0
Oceania, New Zealand and other Small Countries of Pacific	6,500	6.5

Source: Water Distribution on Earth, Wikipedia (https://en.wikipedia.org/wiki/Water_distribution_on_Earth).

Regional Disparities in the Distribution of Rain

Nature has not equitably distributed its resources of fresh water through rain to the different regions of the landmass available for human habitation. Almost half of the global receipts of renewable fresh water in the form of rains are earmarked for the two American continents. The Amazon and Orinoco river basins in South America together provide 15 per cent of the global runoff of fresh water into the seas, against a total of 27.6 per cent from South America, i.e. only 12.6 per cent is contributed by the rest of the rivers in the continent. Yet, one of the driest lands on earth is found in Chile.

The Amazon eco-region and the rain forests of Brazil are important for global climate control because they contribute water to the world through three kinds of 'rivers': (i) most obviously through surface river discharge, (ii) water vapour in atmosphere through transpiration from the rain forests and (iii) discharge of water in the ocean bed through the recently discovered subterranean river, (named Rio Hamza) running underground, almost under the Amazon river.

Brazilian geo-scientists announced in the year 2011 that Rio Hamza runs some 4,000 m below the Amazon river. It is about 5,000 km long and flows west to east. The river bed of this astonishing river is even bigger than Amazon in terms of width. Rio Hamza's width ranges between 200 km and 400 km as opposed to Amazon's 1 km and 100 km. Some geologists regard the bed of Rio Hamza as being made up of porous rocks through which the 'river' moves. Compared to Amazon's 1,33,000 cu. m per second of water flow, Hamza has a water flow of 3,900 cu. m per second; but it flows at a speed of less than 1 m per hour, against Amazon's flow at a speed of 5 m per second. Rio Hamza opens directly into the depths of Atlantic. Scientists believe that the relatively low salinity of water where Amazon meets the Atlantic is probably due to Rio Hamza, also diluting the salt content of the sea water.

But Asia, with almost half the world's population, is treated shabbily by Nature as it receives just a third of the total global rainfall. Again, regional disparities existing within a continent are masked in statistical presentation of data. Wet and dry areas may exist at the same time on a continent or within a country but this situation remains masked in terms continent-wise data. To illustrate the point, the following may now be pointed out:

- Of the total contribution from the Asian continent, more than half of it is contributed by the basins of Ganga-Brahmaputra, Irrawaddy and Mekong. But the lack of rainfall in vast regions of Middle East or the Thar and Gobi deserts is not revealed.
- Even in well watered continent of North America, there are areas like Texas and California that are extremely short of water.
- Despite the huge runoff coming from South and South-East Asia, water availability in parts of Bangladesh and Baluchistan in Pakistan is very poor.

- The data of Asia masks the fact that Yenisei river basin is the second largest river basin in the world after the Amazon basin but poorly inhabited.
- The data for India masks the very dry regions of Maharashtra and Karnataka.

As the world faces increasingly greater scarcities of water, all countries of the world are making efforts to retain as much of the rainfall runoff as possible by making reservoirs in the flow path of the rivers in their own territories.

Habitat of Fresh Water in Ice and Snow

Water in the form of ice and snow is in its purest form on the surface of the Earth. It does not contain any salt; not even when it separates from sea water in the Antarctic Ocean or indeed anywhere else when the temperature of the saline water drops to slightly less than 0°C. However, if the temperature of saline water drops considerably below 0°C, the quantity of fresh water ice would keep increasing until only salt in solid state is left. It is estimated that a staggering two-thirds of total presence of fresh water on the surface of the earth is in the form of ice. Most of it is present on the icy continent of Antarctica and in Greenland. It is also found in Iceland and in the upper regions of Himalayas, the Alps, the Rockies and the Andes mountains. A small amount is also locked in the permafrost beds.

The Antarctica ice sheet is the single largest ice sheet on planet Earth spread over an area of 14 million sq. km or a volume of 26.5 million cu. km. An idea of its physical enormity can be had by appreciating the fact that were this quantity of ice to melt down completely, it would raise the level of oceans by 70 m. The Antarctica continent is totally uninhabited. Thus, there is no human activity to destroy its natural ecology, except that being caused by visiting teams of scientists. The security and stability of its ecosystem is, therefore, the responsibility of all nations.

Almost 80 per cent of Greenland is covered with ice sheets. The weight of the ice sheets is so much that it has depressed the central land area to form a deep basin, having a depth of 300 m. Were the Greenland ice sheet to melt away completely, the world sea level would rise by more than 7 m on its account.

The ice sheets are formed from thick snow beds which get compacted under their own loads. The sheets of ice may move down the slopes under gravity the form of glaciers (or rivers of ice) until they meet the sea where they may form vast floating ice shelves; the shelves eventually carve out into icebergs. The icebergs move randomly with the sea currents and waves; gradually melting and eventually disappearing in the relatively warmer seas. The pristine water of the Antarctica ice becomes saline once again.

The floating ice shelves act as moving rocks in the sea and have tremendous kinetic energy and solidity to shatter any object that strikes them. Ship captains are always wary of their random movement in the vastness of the sea. It is a matter of recent memory that in the year 1915 Titanic, the largest ship ever made, then considered almost invincible, broke into two pieces when hit by the floating ice in its inaugural voyage to United Kingdom across the Atlantic Ocean. The tragedy, no doubt, gave rise to substantial engineering and technological developments.

Impact of Global Warming on Ice Caps

Global warming is due to increased concentration of greenhouse gases (mainly carbon dioxide and methane) in the atmosphere in the last few decades due combustion of fossil coal and petroleum products. Until about five decades back, there was a balance between fresh ice input to the Antarctica continent and that which slid away into the ocean. But today that balance or harmony seems to have been disturbed as a consequence of global warming. According to the scientists of Chile, the temperatures in the region have gone up by about 3°C during the last 50 years. This has resulted in more ice melting than being replaced by fresh inputs. There is an international effort now to negate global warming by reducing the discharge of greenhouse gases into the atmosphere.

Habitat of Water in Permafrost

Apart from the Antarctica at the South Pole, there are vast regions of nearly similar cold lands along the entire Arctic Ocean, around the North Pole where low temperatures around most of the year keep the ground permanently frozen. The frozen ground bed is called 'permafrost'. In the long and dark winters, the temperatures may average around minus 15°C, sometimes dipping as low as minus 30°C. During summers, the temperatures rise to 3°C–12°C. This results in the thawing of ice in the top layers of permafrost bed, leaving the ground very soggy. With water in all spaces in the deeper stratum of permafrost being solid ice, which acts as an impervious bed and does not allow the melted/thawed water to permeate downwards through the bed. The result is that top layers of the permafrost bed become marshy, and are covered with bogs, lakes and streams. This is the so-called Tundra region. There are vast areas of Tundras in northern Russia and Canada, spreading over an area of 1,15,63,300 sq. km. Tundra lands remain covered with snow for much of the year.

Rainfall in the Tundra regions is small at 15–25 cm/year level, close to that of tropical deserts. Continued spell of darkness for months together and harsh winter is not conducive for the growth of trees. In contrast with the deserts, marshes develop here because there is practically no loss of water through transpiration, evaporation and infiltration. Yet there is difference in the behaviour of snow/ice in the Arctic with that on the Antarctica where swamps are not formed. This is because the Arctic region is surrounded by huge landmass which gets quickly warmed up even in the very limited summer months. And Antarctica is surrounded by oceans which act as a sink for the solar thermal energy. Permafrost also occurs in the upper reaches of the Himalayas and the Alps.

It may now be stated that as we dig deeper into the permafrost bed, the temperature deeper down begins to rise once again due to the internal heat of the Earth, limiting the thickness of the permafrost bed. This means that at either boundaries of the permafrost bed, the prevailing temperatures are > 0°C.

In terms of its internationally accepted definition, a bed qualifies to be called a bed of permafrost when it remains frozen continuously for a period of two yeas at a stretch. Generally, such beds occur about 50 cm below the surface and its bed thickness could extend to about 1 m deeper inside. The first 50 cm depth below the surface may not always remain below the freezing temperature due to climate related changes in temperatures. Though it is obvious, it may be reiterated that water in the inter-rock pieces or inter-aggregate spaces or pores within the permafrost bed always remains in frozen state, but that in the stratum below is always in its liquid form but in the region above, it changes its form with the local weather.

In the stratum above the permafrost, the physical state of water could either be liquid or frozen, depending on the surface temperatures. The roots of all surface vegetation, including trees, can penetrate only into this bed and bacteria also survive in this bed. This organically active bed is called the 'active zone'.

Normally, the beds are physically stable. In summer months, ice partially thaws out in the active zone, some water oozes out to contribute to the summer water flow in the Himalayan rivers. The vacant spaces thus created in the active bed are filled up again by fresh water from the rains. The water freezes again in winters. This cycle has been going on for eons together undisturbed.

It may now be stated that when water freezes into ice, there is almost 9 per cent increase in its volume. In practical terms it means that when water present in a stratum freezes, its grip on the fragmented rock pieces in the bed is so tight that the bed behaves as a solid composite mass. This means that even if the permafrost bed is resting on the slopes of the valleys, the frozen mass remains anchored to the rocks and does not slide down. This situation may be disturbed by extraneous events involving seismic shocks or by thawing of binding ice cement. When the latter happens, the cementing force of ice is gone. An attempt is made to explain the emerging situation with the help of Figure 1.

Figure 1

The diagram on the left is a vertical section showing the permafrost bed which is capped by active layer supporting vegetation; the diagram an right attempts to show land subsidence on account of global warming, causing instability of the active layer.

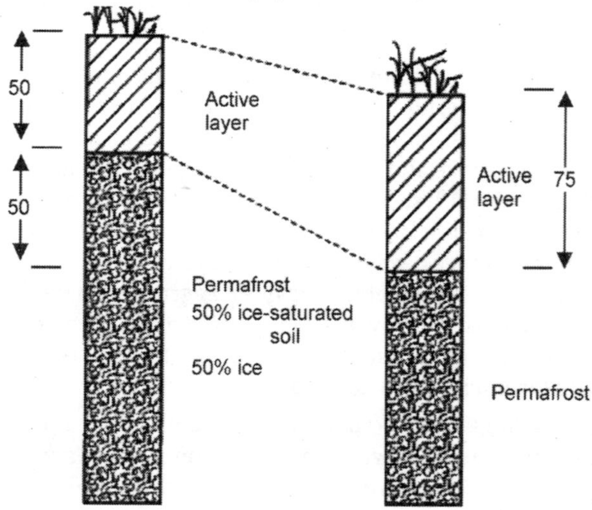

Consequences of Global Warming on Permafrost

For the purpose of explaining the scenario, the section on the left in Figure 1 shows the active zone (for the purpose of illustrating the situation assume the thickness of the bed as 50 cm in which 50 per cent of the volume is occupied with frozen water. Assume further that the active zone has enlarged itself by 50 cm by extending into the permafrost bed on account of global warming. Under this scenario, thawed water looses the tight grip of the erstwhile ice on the soil and rock pieces in the extended active zone. Since water cannot infiltrate down through the residual permafrost bed, it remains in situ in the bed. But the bed has become highly unstable. The section on the right in Figure 1, however assumes that the water has oozed out completely. The bed, devoid of its water, with 50 per cent vacant space becomes unstable. Its collapse could be triggered by slightest seismic activity or excessive rains. When it happens, it does so catastrophically; large pieces of stones and other detritus in the extended active layer roll down the valleys, devastating all that comes in their way.

Such catastrophic events have begun to occur with increasing frequency. The tragedy which occurred in recent past and flooded the Kedarnath holy shrine is largely thought to have been caused by such an act. The magnitude of the damage caused by heavy rains was accentuated by the release of large quantities of water locked in the beds affected by global warming.

Habitat of Water in Lakes and Ponds

Lakes are naturally formed, usually bowl-shaped depressions in the land surface varying in size from that of ponds to those containing vast quantities of water stretching over a few thousand sq. km of area and typically deep. They are usually located in remote areas which may not be easily accessible. Their waters may either be fresh or saline. The source of fresh water is almost always rain water flowing into them through rivers and/or rain water runoff from their bowl-shaped basins, at the deep of which they are located.

Most of the lakes were formed several thousand years ago through tectonic and geological activities. There are five prominent methods of their formation:

1. Some of those high up in the Himalayas are formed by the 'glacial scour' building a 'morain dam'; or through the slow movement of massive amounts of ice during and after the ice ages produced depressions in the land surface which were later filled by water; examples of the latter are the North American Great lakes, lakes in the Lake District (UK), lakes in Scandinavia and Argentina or some lakes in the Himalayas. The weight of massive ice sheets on

Iceland has caused the topography of the island country, Iceland, to take the shape of a dish, i.e. with depression in the central zones.

2. Tectonic movements in which slow movements of the Earth's crust produced depressions over times, which were later filled by water. Lake Tikicaca on Andes Mountains in South America or Lake Nakaru in Kenya are examples of this type of lake formation. Another good example is the Lake Baikal in Russia, which is the largest fresh water lake in the world, and holds nearly 20 per cent of the world's liquid fresh water.
3. When eruption of volcano vomits out magma in such large amounts, the emptied magma chamber underneath collapses forming a large indentation (often called caldera) which fills up with water later. Lake Toba in Indonesia is the largest crater lake in the world. Another good example is the Crater Lake on Mount Mazane in Africa.
4. Impact Crater Lake created in the indentation made when an extra-terrestrial body (an asteroid or a meteor) strikes the Earth. Lonar Crater Lake in Buldhana district, Maharashtra is the best known example of such a lake, being the largest in its class in the world. Its water is saline and alkaline. It is oval in shape with a mean diameter of about 1.2 km and meteor crater with a diameter of about 1.8 km and has an average depth of about 150 m. It was created about 50,000 years ago and finds mention in ancient Indian scriptures such as Skanda Puran or Padma Puran.
5. There are other slow acting geological mechanisms, such as chemical weathering, through which surface ponds/lakes can form. They are then often called 'sinkholes'. A sinkhole is, therefore, an area of ground that has no natural external surface drainage. When it rains, all of the water stays inside the sinkhole and typically drains into the sub-surface.

The formation of sink holes involves some form of collapse of the surface which may occur when the sub-surface strata is made of a limestone/carbonate rock bed or salt beds, or rocks that can naturally be dissolved by ground water circulating through them. When the percolating acidic rain water reaches the rock, it reacts slowly with the bed and may create a network of cavities and voids as a consequence of which spaces and caverns develop underground rains become acidic due to rain drops absorbing carbon dioxide from the atmosphere; when they react with decaying vegetation, the water becomes slightly more acidic). As the process continues, the voids and cracks are enlarged. The land usually stays intact until the underground spaces become too big to support the weight of the land above them. The spaces then collapse suddenly forming a depression on the land surface. A sinkhole is, thus, formed. Such sinkholes form over geological scale of time and are found world wide. Typically, sinkholes form so slowly that little change in surface morphology is noticeable within a finite frame of time. They vary in size up to 600 m in diameter and depth. Some sinkholes are shaped like shallow bowls or saucers whereas others have vertical walls; some hold water and form natural ponds.

Geologically, there are three types of sinkhole (or Karst), lakes: (i) they may occur independently, (ii) may occur on a river path, i.e. with inlets from and outlets into streams/rivers and (iii) lakes formed in conjunction with a sink hole in which the sink hole plays a predominant role.

Network of Sinkhole/Karst Lakes and Karst Topography

About 10 per cent of the earth's land surface consists of soluble rocks such as limestone, dolomite and gypsum which can be easily dissolved by the weak solution of carbonic acid found in underground water. When carbonate rocks interact with weakly acidified underground water, the water dissolves the carbonates to form a characteristic topography – an amalgamation of caves, underground channels, and a rough and bumpy ground surface. This is called the Karst topography after the name of the plateau region in eastern Italy and western Slovenia. The Karst underground drainage systems are characterized by channels inter-connecting caverns/caves and sinkholes. The caverns/caves are susceptible to collapse from the surface, specially when emptied of their waters by excessive pumping out of ground water on account of the load of the overburden. Inside Karst caves, one might find a wide range of stalactites (those structures which hang from the ceilings of the caverns) and stalagmites (which grow on the base). These structures are created naturally when salts carried by the dripping water are deposited slowly over thousands of years.

A stalagmite thus formed, i.e. that which is created naturally, has taken the shape of *shivalinga* (manifestation of Lord Shiva) and has become a place of great veneration at the village, Yana, located in the forests of Uttara Kannada district of Karnataka in the Sahyadri ranges of the Western Ghats which geologically has the Karst type of rock formations. Yana has become a well known pilgrimage centre because of the *Swayambhu* (or 'self-manifested' *shivalinga* or that which is created by its own accord). Since water continuously drips from the roof over the stalagmite (*linga*), it adds to the sanctity of the place.

All types of limestone strata are not suitable for the formation of karst topography. The essential conditions for the development of karst topography are:

- The limestone deposit should be massive, thickly bedded and non-porous. If the rock itself is porous, the infiltrating water will pass through the massive body of the rock, dissolving away the soluble salts and thereby rendering the stratum hollow and weak and susceptible to collapse.
- The limestone stratum should geologically be heavily fractured and faulted so that infiltration of rain water occurs through the spaces at the joints and not through the main mass of the deposit. The water now effectively widens the spaces at the joints by dissolving away the soluble minerals thereby resulting in the formation of caverns and drainage channels.
- The carbonate stratum should be close underneath the surface and be above the ground water table so that rain water may easily infiltrate and its drainage may occur through sub-terranean channels in the stratum.
- There should be adequate rainfall in the region to enable the sub-terranean flow to occur.

In India, extensive Karst topography has not developed because (a) most of the Deccan Plateau is formed through massive igneous eruptions limestones, (b) most of north India is constituted by alluvial soil and (c) limestone strata (say, of the Vindhyan ranges) are buried deep under thick covers of sandstones and shale. Karst topography is restricted to a few places in Jammu & Kashmir, Sahasradhara, near Dehradun (Uttranchal), Panchmarhi, (Madhya Pradesh), Bastar and Raipur (Chhattisgarh), coastal areas near Vishakhapatnam and a few places in southern India. One such place is Yana, as mentioned earlier.

There were many Karst type of ponds in Madhya Pradesh and Chhattisgarh states as the geological conditions there were appropriate for their formation, i.e. exposed, faulted beds of cavernous limestones and sandstones with undulating surface morphological features, having gentle radial slopes. Many of the ponds have been destroyed by urbanization and other anthropogenic activities. Excessive exploitation of ground water through deep bore wells has only accelerated their 'killing' because of enforced induced flow of surface rain water from the shallow surface ponds to deeper ground water aquifers.

These carbonate rocks are of strategic importance for local communities as they provide the main water source for irrigation and domestic use. It may be pointed out that a significant part of aquifer recharge occurs as point-recharge through sinkholes and ground water flow is channelized by conduit networks which sometimes emerge as Karst springs. Intensive quarrying for cement production and building stones is irreparably damaging the point recharge locations, besides the deepening of water table as a consequence of heavy withdrawals of ground water through deep bore wells.

The country must explore alternatives to cement produced with limestone as the primary ingredient. At present, cement companies may be mining Karst type of deposits of limestone extensively and thus destroying important resource of underground water. The limestone mining activities in Chhattisgarh state are known to have already created shortages of potable and irrigation water in few districts (*ne* Baloda Bazar).

Many Karst lakes/ponds in Eastern Europe, Slovenia and Croatia were destroyed or killed in the post World War II socio-economic reconstruction programmes but are now being revived with international cooperation.

Classic Examples of Sinkholes in the World

The form of some sinkholes in the world has made them classic examples of natural geological architecture. Amongst them are:

1. The Cave of Swallows in Mexico: It is over 370 m deep and the largest known cave shaft in the world.
2. Red Lake Sinkhole: Red because of the reddish cliffs that surround the sides of the sinkhole – is an enormous hole that contains a lake. The vertical walls that line the sinkhole are about 290 m in height; however, the lake stretches much deeper than this. With a total explored depth of approximately 530 m and with a volume of roughly 25–30 million cu. m, the Red Lake is the third largest sinkhole in the world.
3. The Great Blue Hole in Belize: The Great Blue Hole is an enormous submarine sinkhole that is located off the coast of Belize. The deep blue colour may look a little intimidating, but the colour of the hole

is merely a contrast between the massive pit and the lighter, white carbonate sands of the surrounding sea floor. Scientists estimate that this formation began to take shape (creating a deep depression in the earth) some 1,53,000 years ago with the collapse of an ancient cave/cavern system, the top of which was worn away by erosion and other weathering processes. It is classified as a World Heritage Site by the United Nations.

Underground Lakes

When the large basin like voids formed through the process of removal of soluble salts from sedimentary strata are filled with the percolating water, an underground lake is created. There are several underground lakes in the world. Some of them have great potential for tourism because of their novel features.

- God's Gift River in Jim Corbett National Park in India.
- Seegrotte, Austria
- Cenote Angelita, Mexico; located at a depth of 30 m; a lake having both fresh and salty water;
- Hamilton Pool, USA. It is one of the most unusual lakes in the world. It is both an underground lake and an island that comes to the surface under the arch stone grotto. The lake itself was formed thousands of years ago after the collapse of the limestone dome of an underground river because of the massive erosion. A grotto is any type of natural cave that is associated with modern, historic, or prehistoric use by humans.
- Lake Banff is located on the territory of the Canadian National Park near the town of Banff. This underground lake flows through dozens of the little caves and is fed by a small waterfall.
- Cheddar Gorge is a part of the largest canyon in Somerset, UK. It became famous because a complete skeleton of a man, now called the 'Cheddar Man', was found there. Scientists estimate that the man had lived over 9,000 years ago! In 1997, a school teacher had his DNA matched to that of the Cheddar Man, proving that generation after generation his family had lived there. The lake is also famous for its stalactites and stalagmites of various shapes and sizes in the caves.
- Unique Melissani cave and an underground river is located on one of the tiniest Greece's islands. According to the legend it was once inhabited by the nymphs, who were thought to be the spiritual defenders of the country. The high walls of the cave that

now 'protect' the turquoise waters of Lake Garda from the heat and wind are believed to remain the blessing of the nymphs that left the land by the will of Gods.

- Kungurskaya ice cave is one of the largest limestone caves in the world and one of the most popular attractions of the Ural. Kungurskaya lake is the largest underground lake of Russia.
- The Gruta do Lago Azul (or the 'Blue Lake Cave'), Bonito, is known for its underwater caves with lakes with large number of stalactites and stalagmites in caves.

Morain Lakes

Glaciers, or ice rivers, move along natural gradients. As they glide along at speeds typically around 1 m/day, they scratch, scour or gouge the surface layers of their beds and carry along the fragmented pieces of the bed rocks. This phenomenon has been happening for thousands of years and has played an important role in sculpting the mountain landscapes. Most lakes in the world occupy basins scoured out by glaciers.

Glaciers leave their load of fragmented rocks just where the ice begins to lose its grip on them on account of its thawing. At places where the dropped pieces of rocks could assemble to take the shape of unconsolidated wall, the flow of water is obstructed and a lake is created in the upstream direction, much in the manner a dam does on a river. The, thus, formed water body is called 'morain lake' and the dam structure, the 'morain dam'. Morain lakes are the habitats of water of most pristine quality. The Himalayas are dotted with morain lakes.

Until recently, the glacial flow was as it was centuries ago. International studies, including those by the Indian Space Research Organization, show that it is no longer so. They draw attention to the unsavoury truth that Himalayan glaciers are becoming the victim of global warming. They are retracting at dangerously fast rates, some may have already disappeared and others are on the brink of vanishing. Whilst their melting may in itself be an innocuous phenomenon, it has threatened the perennial flow in many rivers and existence of many morain lakes.

Glacial Lake Outburst Flow (GLOF)

As long as the strength of the morain wall exceeded the hydrostatic pressure of the lake waters, the dam remains stable. This implied that the inflow of water should be balanced by corresponding outflow. In the event of the inflow imperceptibly exceeding the outflow, water would start building up in the morain lake. The moment the hydrostatic pressure exceeded the strength of the weak dam wall, the dam would breach. The breach often occurs catastrophically, without giving any prior warning. When it occurs it causes unprecedented devastation in downstream areas through massive water flows, defined by the words 'glacial lake outburst flow'. Suitable monitoring of the levels of water may be able to blunt the damage caused by devastating floods in downstream, but nothing can be done to reduce the inflow of water in the morain lake except coordinated worldwide efforts to reduce global warming.

Threat to the stability of morain dams also comes from another source on which the humans have no control. Of late the seismic activity in the Himalayas has increased on account of the Indian tectonic plate building pressure on the Eurasian plate.

GLOF is no longer an issue of human imagination as the phenomenon has begun to occur. The attention of the world was focused on GLOF through a devastating outburst of a morain dam in Nepal on 24 August 1954 in the upstreams of river Kosi. The discharge of its waters when it entered the state of Bihar in India suddenly rose to 24,000 cum/s. The ferocity of its flow can just be imagined by visualizing that a 24 km wide, 1 m deep water front is gushing ahead with a velocity of one m/s. In its violently destructive run on the mountain slopes in Nepal, prior to its entry in India, it had completely destroyed a hydro-electric power station, besides causing tragic floods.

Another similar incident occurred in the morain lake, Dig Tsho, formed by the Langmoche glacier when a portion of the morain wall collapsed suddenly, without giving any warning on 4 August 1985; the Dig Tsho lake measures 1,500×300×18 m. Its water holding capacity is 67,00,000 cu. m. The catastrophic incident was triggered by a huge piece of glacier falling into the morain lake.

International studies have unmasked the unpleasant truth that many morain dams in Bhutan and Nepal are in a very fragile state and could break down on the slightest excuse. The morain dam walls can, however, be strengthened through civil engineering intervention. It is a matter of satisfaction that the concerned governments have begun to adopt appropriate measures in this direction. Indeed, the Tsho Ropla morain lake was in a highly threatened state but its morain dam wall was successfully strengthened in 2000 CE through international cooperation. It may be reiterated that major rivers of both Nepal and Bhutan enter India. Thus any untoward incident related to the rivers eventually affects India.

Geological Information about Himalayas and its Rivers Communicated to Society through Mythological Stories

While facts of science are dry information for the common man, they impact on the lives of everyone. Indian sages of yore had developed a unique system of communicating the fruits of their learning to the common man so that the society as a whole benefits from their knowledge. Following this tradition, they interwove the destructive and reconstructive aspects of river Kosi, much of which flows through the geologically tormented Himalayas, through a very fascinating story. The story shows river as the youthful daughter of exceptional charm and beauty of Lord Shiva of Hindu Trilogy. A demon by the name of Renu Sardar developed passionate infatuation towards her and madly pursued the indifferent Kosi. Harassed by the unwelcome advances of the demon, she condescended to accept him only if he could complete the almost impossible task of building a dam across the valley overnight so as to contain her within the mountain ranges. Thus prevent her from spreading the fertile alluvium in the Ganga plains and meeting the river. Happily excited, the demon started building the dam enthusiastically by putting large pieces of rocks together in the fashion of morain dams. Sensing his success, Kosi appealed to her father, Lord Shiva, to save her from the demon. Lord Shiva took the form of a cock to herald the dawn by crowing ahead of the actual dawn. In non-urban India, it is traditional to this day to believe that cocks crow to herald the morning. Hearing the crowing of the cock, Renu Sardar ran away from the construction site only to realize that he has been cheated. In utter revengeful disappointment, the demon announced that he would visit her year after year. Truly, the demon visits her in the form of land slides in her valleys or rolling down of rock pieces in her path. And Kosi in her youthful energetic exuberance overflows them all. And, thus, succeeds in meeting river Ganga.

Man-Made Reservoirs

Small reservoirs began to be constructed some 4,000–5,000 years ago in China, Egypt, and Mesopotamia and by the Sindhu (Indus) valley civilization, primarily to meet the needs of drinking water and for irrigation. Many experts believe that the Maya Civilization of Central America vanished suddenly because of continued droughts over decades and their lack of knowledge of underground water resources.

In India, regions of scarce water availability led the country in devising ways to store water. Water scarce regions of Rajasthan, Gujarat, Maharashtra and Karnataka were foremost in this regard. Temples almost always had water bodies around them. These small reservoirs adequately met the requirements of the then much smaller and widely scattered communities. But they were not adequate to meet the demands of larger populations.

Although not strictly made by Nature, there are certain lakes in India which mimic the naturally made lakes in terms of the topographical features which determine the flow of water into them. Most prominent amongst them is the Bhoj Tal in Bhopal, locally known as the Bara Talaab, with a catchment area of about 360 sq. km. It was built by Raja Bhoj (1,005–1,055) by building an earthen dam across the Kolans river. According to local folk tales, the inflow from 365 tributaries was first directed into a pond on the advice of a monk.

All natural and man-made lakes are under state of acute stress. Water level in Lake Tikicaca receded by 81 cm between the years 2000 and 2009 due to shortening of rainfall in its catchment areas. Lake Baikal is also reported to be shrinking. The Aral Sea in Central Asia has already almost dried up.

In more recent times, large water bodies began to be created by damming a flowing river or by diverting water from a river to an artificial basin, called impoundment. Like lakes, man-made reservoirs range in size from pond like to very large reservoirs. Large reservoirs being built nowadays are called multipurpose reservoirs as they meet more than one need of the society. These needs include municipal and drinking water supplies, agricultural and horticultural irrigation, flood control, generation of hydro electricity, water for steam required for power generation and industrial applications, development of fisheries, etc. Together these activities strengthen the economic structure of the society.

According to a United Nations Environment Programme (UNEP) publication in the year 2000, an estimated 8,00,000 dams were in operation worldwide in 1997. Of these, about 45,000 were large dams, i.e. dam height of 15 m or more above the natural river bed, or a minimum dam height of 10 m and a volume of at least one million cu. m. It was estimated that globally 25 per cent of all water previously flowing into the sea through rivers is now impounded in man-made reservoirs. The UNEP study indicated that most reservoirs were by that time located in the temperate and sub-tropical zones of the world. Construction of new dams and reservoirs has now virtually ceased in North America and Europe but is going on elsewhere in the so-called developing world. It is still going on with feverish zeal in China but has slowed down in India in recent years.

Most rivers have more than one reservoir built on their flow path. Where the river traverses a considerable gradient in its path, a cascade of reservoirs may be built along its path, so that hydro electric power could be repeatedly generated from the same quantity of water. Reservoirs, tunnels and/or canals with lift pumps are now being built to facilitate inter basin transfer of surplus water of one basin to one experiencing water deficiency. One such project connecting the basin of Godavari river with that of water deficient Krishna river basin was successfully completed in September/ October 2015.

Since they have both river-like or lake-like characteristic, reservoirs may be regarded an intermediate type of water body between rivers and natural lakes. Waters in reservoir have greater residence time than rivers but less than that in lakes. In one important respect, the deeper reservoirs are different from lakes as they have a longitudinal physical gradient from the upstream river end to the downstream dam end. Thus, at their upstream river end, the reservoir waters have river-like or riverine characteristics and at the downstream deeper/deepest dam end, lake-like; this end is called the lacustrine zone. The deepest portion of the lakes is invariably towards their centre.

Earlier, scientific studies of reservoirs were mainly concerned with their sediment loading from their drainage basins as the sediment load is a major factor in determining the useful life of a reservoir. Today, however, we are also concerned with the quality of water in lakes and reservoirs in terms of such variables as water clarity and transparency (greater water transparency or clarity indicates that the water is of superior quality), concentration of nutrients (lower concentration indicates better quality), quantity of algae (lower levels indicate better quality), oxygen concentration (higher concentrations are desirable for fisheries), concentration of dissolved minerals (lower amounts are suggestive of better quality). The science which deals with these aspects is called limnology.

Regrettably, untreated sewage water, many waste and/or chemical compounds are being unmindfully discharged into lakes and compounds with impunity. These may kill useful aquatic organisms or may transfer the toxins to irrigated crops. Because of inadequate water purification, the quantity of bacteria, viruses and other organisms in discharged waters are primary cause of waterborne diseases. As it happens, the threat of disease on their account is severe in developing countries of Asia and Africa.

On account of public neglect, many lakes in India came to the brink of destruction in the recent past when the Government of India, with cooperation from the state governments launched science and knowledge based programmes for their resuscitation. It is heartening to find that lakes that were on the verge of extinction were revived and are flourishing once again – among them being the Pushkar Lake, near Ajmer and the world famous lakes of Udaipur.

Fresh Water Habitat in the Atmosphere

Although atmosphere is not a great storehouse of fresh water, it facilitates transfer of water from the oceans to distant places through air currents. Air currents are winds that move in the atmosphere. In this respect, air currents behave like a 'toll free' super express highway. Evaporation and transpiration change water into vapour which ascends higher into the atmosphere under its own buoyancy, and propelled by the rising air currents. Cooler environs up there cause the vapour to condense and take shape as clouds; strong air currents push them in their own direction until the condensed droplets precipitate as drops of rain or snow flakes.

Meteorologists have estimated that at any point of time, the total quantity of water held in the atmosphere is a mere 0.0001 per cent of the Earth's water; it plays the central role in the hydrological cycle of water. Expressing it in another way, the total quantity of water held in the atmosphere is equal to about 12,900 cu. km of water. Were this quantity to rain on the earth's land mass, it would only collect up to 2.5 cm in the rain gauge. The two other important habitats of water – underground and in rivers – are discussed in two separate chapters ahead in the book.

Global Availability of Fresh Water through Rains

Nature has not equitably distributed any of its endowments to different parts of the earth, be they of fresh water – rainfall, rivers, underground – or indeed in the atmosphere or fertility of land. Water being essential for the survival of organic life, human or otherwise, it is amazing to see that organic life survives in driest of lands as Nature has provided them specific biological features that extract required moisture even from the driest of the atmospheres, or the body adjusts its metabolism to the scarcity of water availability. We have occasions to talk about them later in the book.

Its partiality in distribution of fresh water resources, through annually renewable rainfall, to both North and South America become at once obvious from the data given in Table 5 which lists countries with high annual endowments of rainfall. The three Asian countries – China, India and Indonesia – apparently seem well endowed. But the numbers of persons claiming a share in this bounty of Nature are so large as to make a mockery of this data. The fallacy of this data becomes at once obvious when its availability is considered on per capita per annum basis. This is done in Table 6. India rolls considerably down to 132nd position. This situation is putting stress on the water available in rivers. The position in our immediate neighbourhood, in Pakistan and Bangladesh is no better either; only Bhutan emerges amongst the first ten countries well endowed with respect to water.

Some of the more affluent rivers with respect to water flow through more than one country and each of these countries have a claim on its waters. China has unilaterally built several dams on such rivers within its territories to the concern of downstream countries.

Table 5

Total volume of annual rainfall

Country	*In billion cu. m*
Brazil	8,647
Russia	4,508
United States	3,069
Canada	2,902
China	2,840
Colombia	2,360
Indonesia	2,019
India	1,911
Peru	1,894
Venezuela	1,325

Table 6

Annual per capita water availability (arranged in decreasing order)

	Country	*cu. m/capita/year*
1.	Iceland	5,15,152
2.	Guyana	3,38,750
3.	Congo	1,87,050
4.	Surinamme	1,83,673
5.	Papua	1,09,411
6.	Bhutan	1,03,448
7.	Gabon	99,282
8.	Canada	82,485
9.	Solomon Island	79,679
10.	Norway	77,930
11.	India	1,526

Source: Aquastat, Food and Agricultural Organization's Water Resources Database.

Continuously Decreasing Per Capita Availability of Water in India

On economic considerations, the world is seen to be divided between two groups, those of developed and developing countries. The developed countries are mostly outside the tropics, including Russia, Europe, Canada, the USA, Australia and Japan. With their population levels having already stabilized, and as their per capita water availability is adequate in terms of globally accepted norms, they can be complacent in this respect. On the contrary, the situation is one of continuously decreasing availability in developing countries on account of increasing population, even as the demand for water for use in the industry and agriculture is increasing.

There has been unprecedented increase in the population of India, since 1951. The population then was about 36 crore which is projected to grow to 133 crore by 2025 and to over 150 crore by 2050. Along with urbanization and growth of slum areas around metropolitan cities, this has led to continually decreasing per capita availability of fresh water as shown in Table 7. This is because the every year renewable quantity of fresh water has remained the same since the millennium.

Table 7

Decreasing per capita per annum water availability in India since 1951

Year	*Cubic Metres*	*Remarks*
1951	5,200	On the basis of the then population
1991	2,309	On the basis of the then population
2001	1,816	On the basis of the then population
2011	1,545	On the basis of the then population
2025	1,401	On anticipated population basis
2050	1,191	On anticipated population basis

Internationally, water availability at 1,700 cu. m/head/year is regarded satisfactory to lead healthy life, free of stress arising from water shortages. When its drops below this value, life is regarded as 'water stressed'. When the value drops as low as 1,000 cu. m/head/annum, the situation is one of scarcity. India is fast approaching this situation as a nation but many parts are already grappling with water scarcity. We cannot any longer remain complacent as the situation of water scarcity in these areas is masked by its high availability in certain other areas.

We have to bring about paradigm shifts in our approach towards water resources and their conservation, conserve the rain water presently being returned to the sea as 'surface runoff' water or develop economically competitive technologies for desalination of sea water.

Grim Water Situation in India's Immediate Neighbourhood

Table 8 shows the water availability in our immediate neighbourhood. With the exception of Bhutan and China, the situation in Pakistan and Bangladesh is far worse than in India where the population is bursting at its seams. China could also have been in a similar mess but for its rigorous implementation of one child per family for the last few decades. There are not many options available. Either follow the religious zealots and witness your own progeny dying of thirst or approach the water related issues scientifically. The harsh fact is that a country can import food but water in liquid form cannot be imported from water surplus countries of say South America, though international trade in water has been going on in the form of import/export of agricultural or horticultural produce in the form of sugar, wheat, cotton or fruits or manufactured industrial products, like automobiles, air conditioning equipment, domestic grinders, pesticides or even pairs of manufactured jeans.

Table 8

Water availability (per capita per head) in India's neighbourhood

Country	*During 2005–09, in cu. m*	*During 2010–14; in cu. m*
Bhutan	1,05,147	1,03,456
India	1,169	1,155
Pakistan	307	302
Bangladesh	679	671
China	2,083	2,072

Maintain Sanctity of Natural Habitats of Water or Face Destruction

The 20th century will go down in the annals of world history as one in which mankind made huge advances in the realm of science and technology. It will also be known as one which witnessed destruction of natural resources and most dramatic alterations of earth's surface, particularly in the context of water resources.

One such is the destruction of the Aral Sea which until the mid-1950s was the fourth largest inland water body after the Caspian Sea, Lake Superior and the Victoria Lake. Aral Sea in Central Asia, now in the southern part of Kazakhstan and northern parts of Uzbekistan is fed by two fresh water rivers, emanating from the Pamir Mountains, namely Syr Darya from the North and Amu Darya from the South. The huge input of waters from these two rivers along with rainfall (which met only 20% of the amount evaporated) neatly balanced the evaporation of water from the surface of the Aral Sea and thus was the continuity of Aral Sea ensured. In the 1960s, however, then within the realms of Soviet Union, the government decided to divert the waters of the two rivers to irrigate the cotton fields in an effort to make the area as the largest producer of cotton in the world. The area did become the largest producer but a fairly large amount of water was also getting simultaneously soaked into the desert lands; and thus went blatantly wasted away.

The Aral Sea, devoid of fresh inputs of water, but with evaporation occurring unabated, began to shrink or dry up. It took almost 40 years for the Aral Sea to desiccate and now barely 10 per cent of its former self is left. Wrecks of giant fishing boats and abandoned ships could be seen on the land which until a couple of years back was sea. Efforts are now on to revive the Aral Sea.

As the sea shrank, the pesticides and the insecticides which were earlier washed into the sea with river waters got concentrated to a level which killed the fish altogether; the climate also began to change; rain stopped, grasslands dried up and so did the fresh water lakes that had existed around the sea's edge.

Almost similar is the situation with the Chad Lake, situated at the tri-junction of three African States (Chad, Cameroon and Yobe). Almost seven decades ago, the lake was flourishing with abundant water, but excessive, unilateral withdrawal of water from the rivers, which were feeding it, has reduced the lake to a pitiable state.

The ground water resources in many countries have met similar fate on account of excessive withdrawals to irrigate farms for the production of wheat, sugar and cotton for export to earn hard foreign currency in order to 'boost' their national economy.

Emergence of New Concepts on Water as a Tradeable Commodity

Water is essential for life, so are inanimate objects for the sustenance of the modern style of high quality life. Both need water. Water, therefore, exhibits a kind of duality in its interface with mankind, like the two sides of the coin; obverse (heads or front) and reverse (tails/back). Distinguishing the obverse from the reverse is not always easy in the modern materialistic society.

A new era of global consciousness dawned amongst the enlightened people, specially after the World War II, to provide basic amenities of life (clean drinking water, water for sanitation and to provide food security) to the economically emaciated and socially deprived people in the world. They were mostly residents of countries which came to be known as the Third World countries. As the 20th century advanced, universal usage of the technology of instant mass communication reduced the vast dimensions of the world into a 'global village'. Live scenes of people living under pathetic sub-human conditions could be seen in any corner of the world on their domestic TV screens. They added poignancy to the zealous efforts of the enlightened intelligentsia to provide the deprived people elemental necessities - drinking water, food and sanitation - for dignified human existence. Conscious efforts to translate the lofty ideals into reality, raised consumption of water several folds. Some of these countries are located in the Sub-Sahara region of Africa. Water has always been central in all facets of human activities and will always remain so. There is no alternative to water.

Many of these countries in the continent of Africa and South-West Asia are in themselves poorly endowed by Nature in terms of renewable resources of fresh water. It is a reality that the density of population is more in regions where surface water is easily available, but population has also spread away to water scarce regions as well. Thus both water and people are unevenly distributed even within a country. This is also true on global scale. Therefore, different countries and different regions within a country experience varying degrees of water stress.

In some developing countries, the perpetual shortages of water and food resources have become more acute with population explosion which resulted from improved healthcare. Some of these countries which were denied surface fresh water resources by Nature, but were well endowed with deposits of petroleum, began to exploit their ground water resources heavily (supplementing them with sea water desalination on a large scale) to meet their increasing demands. Others, like India and China, otherwise well endowed with renewable rain water resources used all available possibilities for this purpose – increasing storage capacity for rain water by building dams and reservoirs (where the terrain permitted) along their course of movement to the sea and ground water. These were ostensibly meant to increase farm production to cut down on imports of wheat etc. through modern irrigation based cultivation, but the scarce resources of water were also used to produce water guzzling crops of sugar and cotton, not only to meet domestic demand but also to export to earn hard currency to balance their imports through exports. Even the technologically highly advanced country – the USA – otherwise with very well endowment of natural fresh water on national basis, began using its underground resources relentlessly and its fruit produce could be found almost anywhere in the world.

In their explosive zeal to expeditiously raise the standards of living of the people, some of the developing countries, notably China and India, took recourse to massive industrialization by establishing heavy metallurgical, chemical and manufacturing industries. This not only increased the demand for water but added another dimension to the issues arising out of water scarcity. The industries discharged their waste water, often laden with toxic chemical, into the surface water resources or into the ground which eventually irreparably polluted the ground water resources. This aspect of industrialization, unwittingly polluting fresh water began to exploit its underground water resources specifically in its rain deficit states. In many cases this led to over exploitation of ground water, and consequent drying up of these resources. The health and sustainability of water resources got eclipsed by the urge for expeditious development.

Even as the bounty of Nature in the form of rains, and other natural resources of water – rivers and lakes on surface or underground water – was not equitably distributed across the world, a bad situation became worse for some, when countries in the upstream path of multi-nation flowing rivers began to impound their waters by building dams and reservoirs within their territories, unilaterally. Others, particularly the desert bound countries of South Asia and Northern Africa; the petro-dollar affluent countries met their increasing demands for water by tapping deep underground aquifers, running beneath a number of countries.

With petro dollars, they had access to almost anything the so-called Developed World had to offer in terms of food, fast moving consumer goods or other white goods and luxury items like automobiles, air conditioners, amusement parks, etc. Products which are heavily dependent on abundant water availability such as milk of Swiss cows, beef from New Zealand or Californian fruits were readily available in Saudi Arabia, Libya, and Gulf countries. Even markets in fruit rich India, began to receive fruits grown in United States, China, Israel, etc. even dates from Saudi Arabia.

When goods and services are traded commercially, the benefits of technological advances are available in all parts of the world, if one has the means to pay for them. Thus, models of latest automobiles are available in Dubai where establishment of the industry is not at all otherwise viable. This trade model of costing all ingredients which made the product, into its market price was not available to totally water dependent agricultural produce. It favoured profitable export of industrial products but for agricultural produce the cost of water input was/is not invariably taken into account. The world needed a costing model for agricultural products which could quantify the amount of water consumed in the production of a unit weight of the agricultural produce so that cost of water is taken into consideration in determining the cost of the product. When this is done, water-based unprocessed agricultural produce would become financially as rewarding as the industry-based production of packaged and processed food or any other manufactured product. This is important because 70 per cent of the world's fresh water is used for irrigation but the food processing industries are located elsewhere. Thus, unprocessed farm produce of potato may be selling at a retail price of ₹ 20/kg but processed potato wafers for ₹ 300/kg or sugar mills may be buying the sugarcane at the so-called 'fair and remunerative price, FRP' at ₹ 230/quintal (December 2015) and selling sugar at ex-mill price of ₹ 2,300/quintal.

Water, like air, is available to the industry at low or nominal cost. Wherever the industry pays for the water it uses for its processes and products, the price is far below its real economic cost. The government subsidises the cost of water heavily by building dams, canal and distribution infrastructure. These costs are not charged to the account of the industrial user. As a result, there is no incentive on large or industrial users of water to economize its usage or to prevent discharge of their polluted waters into water bodies in public domain. The consequences of this irrational approach towards water have now begun to show themselves in the form of exhaustion of ground water, pollution of surface water bodies and even of underground resources.

The close relationship existing between water and agricultural or horticultural produce is easily appreciated. Similarly, any industrial process/product requires availability of water in direct or indirect form – water in thermal/hydro-electric or nuclear power, water in mining, metallurgical or manufacturing industries, etc. Effectively, it means that when a country is importing food items, or beverages or manufactured engineering products, it is importing water indirectly, because without water the production of the manufactured goods in the country would not have been possible. We could, indeed, regard the transaction as indirect trading in water. In terms of this concept, an automobile is as much 'packaged water' as is any agricultural or horticultural produce. The need to treat all natural and industrial products, process and services in terms of their water equivalent is increasingly being realized and has given rise to the concept that any agricultural or horticultural produce or engineering product or human activity has a typical 'water footprint' for itself.

Nations may squabble with one another on distribution of water of a river basin but such squabbles for import or export of industrial products are unknown because their prices are based on economic considerations. But subsidies offered by governments to promote exports of farm products become matter for serious discussions at high governmental levels.

The Concept of Water Footprints

The concept to identify a product in terms of its water footprint was developed in 2002 by A.Y. Hoekstra (at the Institute of Water Resources, now known as the International Institute for Hydraulic and Environmental Engineering, Delft, Netherlands, UNESCO). Water footprint of an agricultural produce is defined as the total volume of water used to produce it and takes into account all stages from preparation of farm land and seeding up to harvesting. According to Professor Hoekstra, the creator of the water footprint concept: 'The interest in the water footprint is rooted in the recognition that human impacts on fresh water systems can ultimately be linked to human consumption, and that issues like water and pollution can be understood and addressed by considering production and supply chains as a whole'.

Water Footprints of Agricultural and Horticultural Products

As water footprints are finding increasingly greater use, they are calculated in accordance with the guidelines laid down in International Standards, ISO 14046:2014. The water footprints for some agricultural produce are given in Table 9. Sometimes it is also called the virtual water content of the product because it represents the amount of water that was used in the production of the crop (or the product).

Table 9

Water footprints of agricultural products in litres/kg

Product	*Water, l/kg*	*Product*	*Water, l/kg*
Cauliflower	280	Rice	2,500
Potato	290	Wheat	1,608
Tomato	214	Maize, global	1,220
Orange juice	1,020	Maize, US	760
Mango, Guava	1,800	Maize, China	1,160
Apple juice	1,140	Maize, Brazil	1,750
Pumpkin	350	Maize, India	2,540
Banana, India	500	Cotton, global	10,000
Banana, global	790	Cotton, US	8,100
Dates, Iran	3,400	Cotton, China	6,000
Dates, Saudi	3,650	Cotton, India	22,500
Dates, global	2,280	Cotton, Pak	9,600
Groundnut, global	3,970	Cotton, Uzbeks	9,200

Cont'd...

...Cont'd

Groundnut, China	1,720	Beef	15,400
Groundnut, India	3,420	Sheep mutton	10,400
Beetroot sugar	920	Pork	6,000
Cane sugar	1,780	Goat mutton	5,500
Tea leaves	8,860	Chicken meat	4,330
Coffee roasted	18,900	Egg	3,300
Biodiesel, cane	2,107	Cow milk	940
Biodiesel, soya	11,400	Plastic	200–250
Biodiesel, corn	2,854	Petrol	1–2.5
Biodiesel, beet	1,188	Car of 1,000 kg weight	3,20,000
		Steel	250

Agricultural Crop Pricing and Water Intensive Crops

In the last four decades, roughly 84 per cent of the total addition to the net irrigated area has come through ground water. The primary cause of over-exploitation has been the rising demand for ground water from agriculture. Further, decisions such as cropping pattern and cropping intensity are taken independent of the ground water availability in most areas.

The High-Level Committee on restructuring of the Food Corporation of India in 2014, observed that even though Minimum Support Prices (MSPs) are currently announced for 23 crops, the effective price support is for wheat and rice. This creates highly skewed incentive structures in favour of wheat and paddy, which are water intensive crops and depend heavily on ground water for their growth. Table 10 below shows the average amount of water (in cu. m/tonne) needed to grow different crops. It indicates India's efficiency in the usage of water for agriculture as compared to other countries. As can be seen, India uses almost twice the amount of water to grow crops as compared to China and United States.

Table 10

Water use for crop production in different countries (in cu. m/tonne)

	Brazil	*India*	*China*	*United States*
Rice	3,082	2,800	1,321	1,275
Sugarcane	155	159	117	103
Wheat	1,616	1,654	690	849
Cotton	2,777	8,264	1,419	2,535

Source: National Water Footprint Account, UNESCO-Institute for Water Education, May 2011.

The High-Level Committee suggested that cropping pattern needed to be diversified by providing better price support for pulses and oilseeds. This would also offer incentive the production of these foodgrains. Other experts have suggested the use of demand management in agriculture to address excessive ground water exploitation. This will reduce dependence of agriculture on ground water. These measures include: dry-season crop planning for a specific area depending on the aquifer type, ground water.

Use Water Rationally, Avoid Irrational Use of Water for Unsustainable Production of Crops or Farm Products

We must seriously ponder over promoting rational and scientific uses of water so that the available water resource is gainfully utilized only for agricultural and farm productions which are ecologically friendly. In this context, the production of water guzzling crops, such as sugar cane, cotton or paddy for exports be restricted, through global agreements under the authority of United Nations, to those areas/countries where these crops are wholly dependent on rainwater (the so called green water). Some countries are currently producing these crops for exports by irrigating them with ground water, (i.e. the so called 'blue water') though their native population is struggling with acute shortages of potable water. Pumping out of large quantities of ground water is additionally resulting in ecological disturbance. Their exports should be discouraged by the international community. There is precedent to this suggestion as many countries do not import products from other countries where child labour is employed for their production. We may now illustrate this point with a few examples.

1. Flushed with the availability of the then abundant supplies of ground water, Saudi Arabia promoted the cultivation of wheat. In fact, so extensive was the cultivation of wheat that Saudi Arabia had become the sixth largest exporter of wheat in the first decade of this century. However, massive withdrawal of underground waters led to complete depletion of their deep aquifers. The cultivation of wheat in Saudi Arabia has since been banned through government notification. Earlier, the government had decided to completely phase out wheat production by 2016.
2. It is a matter of common understanding that the natural habitats for cows are the lands where climate is moderate and with plenty of water for the growth of grass and cultivation of other green fodder. Climatically, therefore, the dry desert lands where

ambient temperatures in day time soar to over 50°C provide hostile environment for their survival. Sensing the political inclinations of the Arab rulers in the post-1970 years, after the unilateral hike in prices of petroleum, to provide food security to their people at any cost, innovative marketing experts began to proclaim that dairy industry can be established almost anywhere in the world, provided one has the will and the finances to do it. They convinced the governments of some Arab countries that as long as water and feed could be sourced, dairy operations could be carried out almost anywhere in the world, including in dry Arab lands where not a blade of grass grew naturally.

It is generally understood that the dairy industry requires inputs of three different kinds: (i) cow breed/s suitable for the local climate, and if the natural climate is not favourable, it can be altered with the help of available climate control technologies, (ii) cattle feed grass and crops, the cultivation of which requires plenty of irrigation, and (iii) plenty of water for the cows. They advanced the logic that grass and other feed crops can either be produced with own ground water resources, then thought to be available in unlimited quantities, or by supplementing it with imported alfalfa feed, either fully or partially. Even where the feed is totally imported, the cows still require enormous quantities of water to yield milk. Dairy experts say that one cow consumes about three litres of water to produce one litre of milk; some of the high milk yielding cow breeds produce 50 litres of milk/day. That equals to 150 litres of water per cow/per day. That world's best dairy cattle breeds – Jersey, Guernsey, Holstein – are native of some of the wet, lush, and green places in Europe, is itself a testament to the fact that to produce milk, one needs a lot of water.

Lush with petro-dollars, Saudi Arabia and other Gulf countries attempted a tryst with Nature by starting dairy industry in the late 1970s. They imported the high milk yielding cow breeds – Holstein Friesian cows – from Europe and mimicked the European climate for them, through climate control technology by effectively cooling down the temperatures in cow sheds to 21°C–23°C from the almost scorching ambient atmospheric temperatures around 50°C. They grew fodder with the help of ground water and also supplemented it with imports.

Despite adverse weather conditions and severe shortage of cultivable land and water resources, integrated dairy farming activity in desert countries began to be carried out on large scale. Saudi Arabia and other desert countries in the region began producing and exporting enormous quantities of milk and other dairy products. Once heralded as exemplary integrated dairy farming in the desert, it soon brought Saudi Arabia to the brink of disaster of unparalleled dimensions. Today, in just about 40 years, Saudi Arabia's underground store of fossil water, estimated at 500 cu. km, is nearly exhausted. It is estimated that four-fifth of the aquifers are now depleted.

Mexico is another country which is using its lean water resources for the dairy farming and has taken a place amongst the major exporters of beef. The fresh water resource position is grim in Mexico.

Faced with the looming crisis for fresh water, Saudi Arabia will stop the cultivation of all kinds of green fodder by about 2019 in order to save water and replace it with imported green feed. The move is aimed at rationalizing water consumption in agricultural areas so that the farmers change their activity to less water intensive crops, such as potato.

Experts at the King Faisal University expressed apprehensions in 2016 that Saudi Arabia's ground water resources will run out in the next 13 years. The fact that per capita water consumption in Saudi Arabia is 265 litres/day, (i.e. almost double of the European Union average) suggests that there is unpardonable wastage of fresh water in the country. Many of the surface water resources in Saudi Arabia have 'turned into sand pits'.

3. News of similar foolhardy actions is coming from other parts of the world as well. The latest comes from Arizona in United States of America. Faced with acute shortage of water in its own country, the Saudi dairy industry is reported to have bought land in the arid Arizona State to grow hay to export the green fodder to Saudi Arabia to sustain their dairy farms. The local population in Arizona is reported to be agitated over the issue as the State itself is faced with drought.

4. Cotton is the most important natural fibre used in the textile industries worldwide and is mostly produced by irrigating the fields by ground water or by diverting rivers to cotton crop lands.

About 53 per cent of the global cotton fields are irrigated, producing 73 per cent of the global cotton production. The water guzzling cotton crop is mainly grown in the water deficient regions USA, Egypt, Uzbekistan, Pakistan, India and China who together produce about 70 per cent of global cotton production. In all these countries, fresh water is already in short supply and extensive use of ground water has caused falling water tables and the quality of underground potable water has been adversely affected on account of ground water resources getting polluted by the ingress of water from the fields laden with the insecticides used in cotton cultivation. Besides, it has also causing serious consequences on the local ecology and environment – somewhere more noticeable than at others. The Aral Sea is the most notorious example of the effects of river water diversion for irrigating the cotton fields. In the period 1960–2000, the Aral Sea in Central Asia lost approximately 60 per cent of its area and 80 per cent of its volume as a result of the annual abstractions of water from the Amu Darya and the Syr Darya – the rivers which feed the Aral Sea – to grow cotton in the desert. It is reported that in China, some 28,000 of its 50,000 rivers and their tributaries have dried up – it is difficult to say how many of these are exclusively due to irrigation of cotton fields or due to prolonged periods of droughts or global warming.

5. The Australian dairy industry stands out in sharp contrast to that of Saudi Arabia. Prior to European settlement, there were no cattles in Australia. Today, Australia has 3 per cent of the world cattle inventory, with India, Brazil and China taking the top three places. However, the beef industry is the largest agricultural enterprise there. The cattle are reared primarily on pasture lands to export beef, mainly to the USA and Japan. The pastures cover an area in excess of 200 million hectares, growing grass and other vegetation as the cattle feed, distributed in a wide range of climates. They include both the high rainfall zones and low rainfall pastoral zones, characterised by less fertile soils. Australia chose the European breeds for its high rainfall zones and the Indian breed for the low rainfall regions. Fresh water is needed to feed both kinds of livestocks and to keep the animals in good health. In the low rainfall pastoral areas, ground water is being withdrawn to meet the requirements of fresh water. In this context, the Australian dairy industry is sustainable.

Water Footprint of Industrial Products

Water is essential for all kinds of industry - metallurgical, mining, petro-chemicals, fertilizers, pharmaceutical, paper, etc. It is also an integral part of all kinds of industries for manufactured products - plastic packaging materials, shirts and other wearing apparels, computers, medicines, etc. Water is intimately connected with the processing of all of them. Therefore, the water foot print of any manufacturing process or product can be determined according to the standardized procedure referred above. Data for only a few of the products is given in Table 9 for the purpose of illustration only.

When we review the situation from published data for individual products/processes, certain harsh realities come to light. For example, it does not immediately strike us that industrial processes involved in the production of one throw-away plastic water bottle of one litre capacity consume as much water. Similarly, water footprint of a shirt is 2,500 litres and that the footprint of a shirt made from synthetics is less than that of cotton. Or that of a pair of jeans is a staggering 8,000 litres. We willingly pay ₹ 15 for a litre of bottled water and on that count the price of a pair of jeans should exceed ₹ 1,20,000. And yet the market price is substantially less. There is some fallacy somewhere. If the world's water resources are threatened today, it is largely because the society has failed to accord to water the economic dignity it deserved.

The water footprint of aluminium foil, taking into all stages of its production from mining for bauxite, production of alumina and electrolytic production of aluminium metal, ending with the production of its foil, is very large. Aluminium foil is being increasingly employed for its one time use for food packaging by affluent families or high-end restaurants; but we remain pathetically immune at this kind of wanton destruction of water in the face of growing scarcities.

Blue, Green and Grey Water in Water Footprints

The water used in farms and industry comes from two basic sources: renewable rain water which is called the 'green water' and the underground water, called the 'blue water'. Together, these components provide a comprehensive picture of water use by delineating the source of water consumed either as rainfall/ground water. The term green water refers to the volume of rain water consumed during the production processes of agricultural and forestry products. It is the total rainwater used in raising the crop, i.e. that lost through evapotranspiration (from fields and plantations) plus the water incorporated into the harvested crop or wood. Consumption refers to the volume of fresh water used and then evaporated or incorporated into a product.

The term blue water refers to the volume of surface and ground water consumed as a result of the production of a good or service. It is the amount of water abstracted from the ground, or surface water that does not return to the catchment from which it was withdrawn.

The industrially polluted waste water is discharged into surface river bodies and it pollutes them too. When mixed with fresh water, the concentration of pollutants in the water body is diluted. The water footprint concept takes into account this aspect too in the utilization pattern of water and has evolved the concept of 'grey water' which is the amount of fresh water required to dilute the polluted water to an extent that the concentration of pollutants is brought down to a level acceptable for use. In other words, grey water is the quantity of fresh water required for the assimilation of pollutants. In this concept, the water use pattern of the industry is sub-divided into three categories – green, blue and grey.

In so far as the use of green water in farms is considered, it may be stated that green water refers to water stored in the unsaturated zone in ground. Agriculture is based both on rain water (green water) that enters the unsaturated zone or the root zone of the plants and on underground aquifer water (blue water) drawn out by water pumps. When agriculture is based on green water, it is sustainable. Global data shows that 70 per cent of the global wheat production is based on green water, 19 per cent on blue water and 11 per cent on grey water. In some parts of the world, most of wheat and maize production is based on blue water. Similarly, sugarcane production in Brazil and Thailand is based on green water, but in Maharashtra state in India, it is largely on blue water. Many sugarcane farmers in the state are in a state of acute distress as the ground water level has receded so far below in the sub-surface ground on account of excessive pumping out that it cannot any longer be tapped.

Water Footprints of Industry and Individuals

Every industry can determine its own water footprint in terms of green, blue and grey waters. When this is done, the water footprint of manufactured paper is very high – ranging between 300–2,600 cu. m/1,000 kg being at the higher end for virgin paper from trees and at the lower end for recycled paper. Any industry should regularly carry out an internal audit to minimize consumption of water. In fact, when a paper industry in Germany determined its own water print, it not only could economize on its water usage but also succeeded in reducing its grey water component through effluent reduction by ensuring its purification and recycling.

Individuals can also determine their own water footprint. When this is done, it is found that, other things being equal, the water footprint of a vegetarian person is less than that of the non-vegetarian; amongst the non-vegetarians themselves, that of chicken meat eaters would be less than that of beef eaters.

National Water Footprints

The water footprint of national consumption is defined as the total amount of fresh water that is used to produce the goods and services consumed by the inhabitants of the nation. The water footprint of national consumption is calculated as 'the total use of domestic water resources plus the gross virtual water imports minus the gross virtual water export.' Water imprint of a country may be inside the country and partly outside the country depending on whether the products are being locally produced or imported.

Mekonnen and Hoekstra (2011) collected and analysed the water consumption data and pattern of many countries of the world assiduously. The global average water footprint per capita was found to be 1385 cubic metre/year/ for the period 1996–2005; of this 27 per cent was for agricultural produce, 22 per cent for meat and 7 per cent for milk. Their data showed wide differences, not only in terms of total consumption of water, but also in terms of its source. It may be recalled that the total consumption is built up by sum of water used from its own surface water resources (green water) plus that drawn from the ground (blue water). Additionally, it also takes into account that imported or exported invisibly in the form of agricultural and farm produce or through manufactured products. Table 11 is prepared from their data, randomly for few countries. The table also incorporates data for grey water component which reflects the state of their concern for pollution of their water resources; higher values are indicative of higher ground water pollution levels. The water footprint of a country depends on (a) the food consumption pattern,

(b) farming practices, (c) consumption of both fast moving and white consumer goods, and (d) the lifestyle; and therefore on the water footprints of the products and services.

Table 11

Average water footprint of national consumption in m per year per capita

Country	*Total Water Footprint*	*Blue Water Footprint*	*Grey Water Footprint*
Canada	2000–2500	250– 350	> 1000
U.S.A	2500–3000	250–350	>1000
Mexico	1500–2000	250–350	216–300
Brazil	2000–2500	50–100	150–216
Argentina	1500–2000	100–152	150–216
Chile	750–1000	153–250	150–216
Australia	2000–2500	250–350	216–300
China	1000–1250	100–153	300–500
India	1000–1250	250–350	150–216
Pakistan	750–1000	350–600	150–216
Indonesia	1000–1200	250–350	150–216
Yemen	750–1000	350–600	5–100
Saudi Arabia	1500–2000	350–600	216–300
Russia	1500–2000	5–50	150–216
Western Europe	1385–1500	50–100	300–500
Portugal	>3000		500–600
Spain	2000–2500		>1000

Following broad conclusions can be drawn from the work of Mekonnen and Hoekestra:

- The per capita water footprints for developing countries vary much more than for industrialized nations; the largest total water footprint for Mongolia is about seven times the smallest total water footprint for DR Congo. At the low end is the Democratic Republic of Congo, with 552 m^3/yr per capita. At the high end are Bolivia (3468 m^3/yr), Niger (3519 m^3/yr) and Mongolia (3775 m^3/yr).
- The blue water per capita footprints for countries in Central and Southwest Asia and North Africa are highest in the world. Consumers in Turkmenistan have the largest blue water footprint of all countries, namely 740 m^3/yr per capita. Other countries with a large blue water footprint are (in descending

order): Iran (589 m^3/yr/cap), the United Arab Emirates (571), Egypt (527), Libya (511), Tajikistan (474), Saudi Arabia (447) and Pakistan (422). The global average blue water footprint of consumption is 153 m^3/yr per capita, which is 11% of the total water footprint.

- The variation in blue water footprint per capita across countries is huge, much larger than the variation in total water footprint per capita. In order to reduce their consumption of blue water, some countries, notably Mexico and Spain, are resorting to reduce it through imports of agricultural produce.
- Industrialized countries have their per capita water footprints in the range of 1250-2850 m^3/yr. The UK, with a water footprint of 1258 m^3/yr, is at the low end of this range, while the USA, with a footprint of 2842 m^3/yr, is at the high end. The differences can be partially explained by differences in consumption pattern. In the USA, for example, average consumption of bovine meat – one of the highly water-intensive commodities – was 43 kg/yr per capita, about 4.5 times the global average, while in the UK this was 18 kg/yr per capita, about two times the global average.

Similarly, where agricultural productivity or yields are low for same input of irrigation water, water footprint for the crop/product would be high. Thus, the very low agricultural yields and associated large water footprints per unit of harvested crop produce in some developing countries explains why they may have a large, a relatively large water footprint of water consumption. Armed with scientific classification of water use, national governments and planning bodies may introduce measures to improve the efficiency of water utilization and to change agricultural practices necessary to protect their water resources from mal-irrigation practices or by switching over to less water demanding crop varieties. The world has to learn a great deal from Israel in this regard which has turned a desert land into a global fruit basket.

A rational approach towards water utilization in agriculture and horticulture as well as in industry holds the key for sustainable development.

Some Facts

The following situations emerge out when data on global and national water footprints was analysed by Hoekstra and his colleagues and help us to understand and appreciate the invisible role played by water.

- In the USA the average water footprint is 2,842 cu. m/year/capita; in China it is 1,071 cu. m/year/capita whereas it is 1,089 for India. The water footprint of a three-member US family is 262 litres per day.
- 20 per cent of the American footprint is external. External footprint of China is only about 10 per cent.
- Japan's water footprint is 1,380 cu. m per year per capita but about 77 per cent of it lies outside the borders of the country.
- Of the total consumption of water in the world in the period 1996–2005, 74 per cent was green, 11 per cent blue and the remaining 15 per cent grey. Agriculture consumed most of it, at 92 per cent.
- External water dependency of all countries together constitutes 22 per cent of the global water footprint. The share of external water footprint, however, varies from country to country. Some European countries, such as Italy, Germany, the United Kingdom and The Netherlands have external water footprints contributing 60–95 to the total water footprint. On the other hand, some countries, such as Chad, Ethiopia, India, Niger, DR Congo, Mali, Argentina, and Sudan have very small external water footprints, smaller than 4 per cent of the total footprint. Countries with a large external water footprint apparently depend upon fresh water resources in other countries. Of this trade, nearly 76 per cent is for agricultural produce, and 12 per cent each for meat and engineering.
- Water footprints of developed industrial countries are more than those of agricultural countries. On scrutiny, it was found that almost 44 per cent of their water footprint is consumed in the production of electricity through the thermal route. In order to conserve water, these countries have launched massive programmes for generation of electricity through alternative but renewable sources such as solar energy and wind energy.

As stated in a UN document, 'water is an astonishingly complex and subtle force in an economy. It is the single constraint on the expansion of every city, and bankers and corporate executives have cited it as the only natural limit to economic growth'. So serious are the political dimensions of the problems arising out of water scarcity that scientists, sociologists, economists and water management gurus in the contemporary world are united in calling upon the governmental leaders to judiciously use water resources, without politics, so that natural resources can be profitably used/shared between water deficient and surplus countries. Many believe that the recent (2015–16) exodus of people from the Middle East seeking shelter in Europe, ostensibly due to raging wars in the Middle East, is really triggered by the shortages of potable water. Drought conditions in Syria and surrounding regions have been prevailing continuingly for the past 6–7 years at a stretch. The problems in Yemen are also believed to be due to water famine in many parts. Do we have to face catastrophes and chaos before we understand that real action has to be taken now to conserve the water resources and restore to them the dignity that is their due.

Inter-state disputes over sharing of river waters are not uncommon in India. Fortunately, the country has the authority of the Supreme Court to decide the issues on scientific merits. The achievements of Israel – a desert-bound country – in harvesting water and its judicious use, inspire the hopes of future.

Sound of Caution from United Nations

'The United Nations warns that water use is growing at twice the rate of population growth. Unless this trend is reversed, two-thirds of the global population will face water "stress" by 2025. In light of this, sharing water fairly and sustainable amongst so many of us is one of the greatest challenges we face in the 21st century.'

Reduce Individual/National Water Footprints or Perish

There are several ways through which the efforts of individuals, societies and governments together can help in the global task of conserving water. They range from obvious and mundane to thought provoking ones involving science and technology. Truly, the list of measures has no limits; only a few of them are listed below:

1. Reduce huge wastage of water in washing bikes, personal cars and public vehicles like trucks, railway transport coaches, etc. Organized sector may consider collection of used water and its recycling.
2. Reduce water consumption in toilets by installing low flow models, install aerators in faucets, minimize personal consumption in bathing and washing clothes through self-innovative methods.
3. Reduce consumption of electricity as production of 1 KWh of electricity through the thermal route requires about 100 litres of water.

4. Avoid giving electricity free or at flat rates to farmers to draw unlimited ground water for flood irrigation; train them to use drip irrigation as far as possible.
5. Avoid wasteful and one time use of energy and water intensive products, such as aluminium foils, throw away plastic spoons and forks and other similar items.
6. Promote recycling of waste municipal waters and multiple use of recycled water in gardening, etc. A study in United States revealed that lawns are the largest irrigated 'crop', mostly using the blue, i.e. ground water.
7. Change personal food and clothing habits, changing from water intensive meats to those with lower water footprints or minimizing the use of water intensive apparels like jeans.
8. Organise opposition to imports of products from countries using blue water in their production or those having considerable grey water component. This is particularly valid for cotton, wheat and sugar.
9. Involve agro scientists in evolving low water footprint varieties of crops or those which can withstand prolonged period of drought or floods.
10. The industry be compelled to discharge water of the same quality as it receives. Large consumers of ground waters be compelled to recharge the ground with as much quantity as withdrawn.

Truly, the list is endless.

References

(All retrieved in January 2017)

Causes, Effects and Types of Sinkholes, Conserve Energy Future http://www.conserve-energy-future.com/causes-effects-and-types-of-sinkholes.php.

Florida Sinkholes, Karst Definition of karst, meaning of karst, Sink ...inspectapedia.com/vision/Karst_Sinkholes.php.

Geologic Controls on the Formation of Florida Sinkhole, Lakeshttps://pubs.usgs.gov/of/2000/0294/report.pdf.

Global Water Resources: http://www.unep.org/training/programmes/Instructor%20Version/Part_2/Activities/Economics_of_Ecosystems/Water/Supplemental/Global_Water_Resources. pdf.

Habitable Planet Unit 8 Water Resources, https://www.learner.org/courses/envsci/unit/pdfs/unit8.pdf.

Hoekstra, A.Y. and Mekonnen, M.M., 2012, The Water Footprint of Humanity, Proceedings of the National Academy of Sciences, 109(9): 3232–3237, pdf.

Mekonnen, M.M. and A.Y. Hoekstra: The green, blue and grey water footprint of crops and derived crop products: Hydrol. Earth Syst. Sci., 15, 1577–1600, 2011 www.hydrol-earth-syst-sci.net/15/1577/2011/ doi:10.5194/hess-15-1577-2011 Hydrology and Earth System Sciences

Mekonnen, M.M. and Hoekstra, A.Y. (2011) The green, blue and grey water footprint of crops and derived crop products, Hydrology and Earth System Sciences, 15(5): 1577-1600. PDF

Mekonnen, M.M. and Hoekstra, A.Y. (2016) Four billion people facing severe water scarcity, Science Advances, 2(2): e1500323. PDF

Moraine Lake – Wikipedia https://en.wikipedia.org/wiki/Moraine_Lake

Sinkhole – Wikipedia: https://en.wikipedia.org/wiki/Sinkhole 6.

The Water Footprint Assessment Manual: A.Y. Hoekstra, A.K. Chapagain, M.M. Aldaya, M.M. Mekonnen | Earthscan, London, 2011

UNEP (2002). Vital Water Graphics: An Overview of the State of the World's Fresh and Marine Waters. http://www.unep.org/vitalwater

Water Footprint Network: http://www.waterfootprint.org/?page=cal/WaterFootprintCalculato

Water Footprints of Key Crops: National Geographic: http://environment.nationalgeographic.com/environment/fresh-water/global-water-footprint/

Water for a Sustainable World: The UN World Water Development Report, 2015: http://unesdoc.unesco.org/images/0023/002318/231823E.pdf

What are sinkholes and how are they caused? | TRT World www.trtworld.com

What Are Sinkholes? – www.livescience.com › Planet Earth

Wisconsin Geological & Natural History Survey – Karst and sinkholes https://wgnhs.uwex.edu/water-environment/karst-sinkholes/

CHAPTER 5

Ground Water

The generic term 'ground water' conventionally qualifies those natural habitats of water that occur within the earthy mass of the earth as are rechargeable by rain water for subsequent use; specifically it does not include its surface habitats. Mankind has been familiar with ground water from times immemorial and has since been using it in ever increasing quantities. It is the second largest source of fresh water. Unlike its habitats on the surface of the earth, ground water does not ordinarily occur as a massive body like a lake or a river. It primarily occurs in porous ground/soil in the so-called 'saturated zone' or in fractured and fissured sediments/rocks/strata. It typically fills up the pores and small spaces which are necessarily created when an infinite number of small particles of irregular shapes are packed together to make the soil/stratum or in the spaces created in a stratum by the washing away of its soluble minerals or at the spaces created at faults and fissures or cracks in the bed rocks through tectonic activities.

It is that part of rain water that has infiltrated through the soil to the water table and is stored there. Underground water is never static; it continues to percolate downwards and flows very slowly in the direction of the slope of the underground stratum of its residence. In other words, it flows like a river, albeit very slowly. Unlike fresh water resources on the surface of the earth, and underground water resources in alluvial soil and other unconsolidated stratum, (which receive fresh inputs through rains), water in some underground aquifers was collected by Nature thousands of years ago. Geologists call them 'fossil aquifers'.

It is not known how mankind became aware of presence of water underground. The first association of mankind with ground water was perhaps with its appearing as springs at few places. Instinctively, he soon realized that the ground had some hidden quantities/stores of water which could be induced to come out and collect itself if deep enough wells are dug into the ground.

Fresh water springs may also occur underground. Such springs discharge their fresh water into the sea. The existence of such springs, off the Gulf coast, was known to the ancient Dilman civilization of the Arabs which had flourished around 540 BCE in the region of present-day Bahrain. The word Bahrain in Arabic means 'twin waters', suggestive of the fact of underground springs of fresh water mixing with the saline sea water. Divers of yore dived into salty sea with empty animal skin bags and came up with the bag filled with fresh water along with a haul of fresh water oysters, inside the bodies of those which were growing the highly prized pearls.

The reachable depth for ground water was initially restricted to the depths from where water could be drawn out by human or animal labour through a bucket and rope, mounted on a pulley. Only in recent times, hand pumps began to be used to draw the water out for domestic applications but soon began to be replaced with electricity driven pumps. Dug wells are the main stay of potable water supplies in many Indian villages. Since the geomorphology of the soil greatly varies across India, the depth at which water is found also varies from just a few metres below the surface in some areas near the rivers to several hundred metres deep in deserts. In the northern Ganga plains, it could be found 2–10 m below the surface, in its southern regions, 10–20 m below or in western Gujarat and Rajasthan, 20–40 m. Ground water now meets a sizeable portion of irrigation needs in those parts of the country which are not within the reach of canals and also in the neighbouring agricultural country, Pakistan.

With the advancement of technology, the dug wells began to be replaced with 'bore wells' and the reachable depth found a quantum increase. During the last couple of decades, however, ground water, stored deep in the bowels of the earth, began to be pumped out from 'drilled wells' with the help of powerful electrically driven water pumps. Deepest drilled wells in the world are/were about 10,000 m deep. Exploration for ground water was conventionally restricted to depths 750/1,000 m below the surface. Modern techniques of exploration for water have made the depth criterion redundant as water is being pumped out from far greater depths. With modern technology, water could be made available literally anywhere in the world, even in the deserts of Saudi Arabia or Libya. As the population of the earth increased and human habitations moved away from being close to traditional resources of surface water, ground water proved to be a great boon. Even the desert lands started blooming with agricultural farms.

Ground water now meets the requirements of drinking water for a sizeable human population. According to the estimates prepared by United Nations in 2006, ground water meets almost 50 per cent of the global needs of water. Global data shows that nearly 75 per cent of the residents of the European Union are dependent on ground water for their needs. It is the main source of water supply in many Indian villages, both for drinking and irrigation.

Significantly, ground water brought about revolutionary changes in agriculture as its use in irrigating farms provided an alternative source of water which could be used as and when the farm needed watering. Agriculture moved away from its total dependence on rain and other sources of surface water.

Ground water has the additional advantage of being made available just where it is required, i.e. without the need for any infra-structure for its distribution. Its other advantage is its pristine quality that can be used without any prior treatment. However, this advantage is soon getting lost as many ground water resources are getting polluted due to the infiltration of highly polluted water, contaminated with pesticides, insecticides and pathogenic bacteria.

It is known from common observation that the primal source of surface waters, including of rivers, is rain, but that it is also the sublime source of ground water and is not immediately perceptible. That close relationship exists between rains and ground water was appreciated from the observation that it is available at shallow depths soon after rains but at deeper depths in summer or in periods of drought. According to hydro-geologists, rain water feeds nearly two-third of shallow depth ground water resources and the rest comes from river beds, ponds and lakes, manmade reservoirs and irrigated fields. These can be recharged by rain water. It is now well appreciated that the discharge from ground water must be balanced by fresh recharge in order that the ground water availability is sustainable; otherwise the yields will decrease and the wells would dry out.

Hydro-geologists, however, classify ground water resources as either 'replenishable', i.e. that where the withdrawals could be replen-ished with fresh water or as 'irreplenishable' where it cannot. Once water in the irreplenishable aquifers is exhausted, the aquifer cannot be reactivated in a short period of time. Questions do, however, arise that whether waters in irreplenishable aquifers are everlasting and could mankind continue to withdraw water from them in unlimited quantities forever. In more recent times, we have learnt the hard way that the ground water resources are not inexhaustible as some of the deepest bore wells in the world, i.e. those in Saudi Arabia, or in California (USA), have dried up, merely after 25–30 years of pumping out of water in huge quantities; but the one that supplied water to Mexico City has begun to dry up now after meeting the water requirements of the city for over 100 years. The position with other ground water deposits in the world is not too happy either. There are other questions too that need answer/s. Why has the quality of water in some of them changed after long continuous pumping out? Or why heavy withdrawals from some wells have adversely affected the quantity of water flowing in the rivers in their neighbourhood.

Infiltration of Water: Permeability of Soil/Stratum

We do know from common experience that rain water or water that has spilled on the surface of the ground infiltrates into its inside, under the influence of gravitational forces. It infiltrates rapidly in a dry ground after summer, but the rate slows down as the ground gets soaked with water on continuing rains or as deeps of the ground are reached. Further, the rate of infiltration varies with the nature of top soil – it is very fast in sandy soil, slightly slower in alluvial soil and slower still in silted or loamy soil; and perhaps none in cemented marble or granite floors.

It is also known that the soil/rock bed through which water is infiltrating down retains a small portion of it in its vacant spaces between its particulates as soil moisture; just as washed cloth remains moist even after squeezing out its water. Thus, at no point of time, the deeper layers of top soil are completely dry. As a proof thereof, the sprouting of new leaves and flowering in summer in trees of mango, *neem (Azadirachta indica)* and *amaltas (Golden shower* or *Casia fistula)* may be cited. It is on this soil moisture that the trees survive in summer months, despite the heavy loss of water occurring through the phenomenon of transpiration from leaves. Rains only follow the flourishing of these trees. Indeed, some of the traditional methods of forecasting the timing and the quality of monsoon rains base their anticipations on the timing of the peak flowerings on these trees and their quality.

The apparently solid looking mass of top soil is constituted by the aggregation and compaction of millions and millions of individual solid particles of very small dimensions and irregular shape. In their three-dimensional packing to make a compact mass of ground soil, voids or interstitial vacant spaces and gaps between neighbouring particles are necessarily created. Water infiltrates into the deeps of the surface through these available spaces.

The complexity of the situation can be perceived through a model; take a transparent bucket and fill it up with spheres of uniform size of a transparent solid material; the interfacial surfaces and interstitial voids could then be easily seen; also note that the voids are not interconnected. The total volume of space occupied by the interstitial voids could be experimentally determined and is called the porosity of the system. The porosity is defined as the percentage of the volume of the pores to the total volume of the porous mass. It gives an idea about the maximum amount of water which the porous mass could hold.

When the experiment is repeated by filling the bucket with progressively smaller spheres of uniform size, it would be noticed that whilst the number of interstitial voids increases, their total volume decreases. Yet the pores are not interconnected, i.e. the assembly is not permeable, though porous. But if the bucket is filled with spheres of variable sizes, many or some of the voids are interconnected and such an assembly is said to have become both porous and permeable. Similar is the situation with soil, i.e. if it is built of particles of smaller and irregular sizes, the voids are smaller but more numerous than when larger spheres are used. When a bed of such a soil holds water in all the pores and spaces, it passes the first test to qualify to host an aquifer but certain other conditions have also got to be satisfied.

Figure 1 shows the manner in which particle size matters in the making of ground mass. In common parlance, a bed with 20 per cent porosity is regarded as one with high holding capacity for water and the one with 5 per cent porosity, poor or small. Well-rounded coarse grained sediments usually have higher porosity but sediments with assorted sizes have lower porosity because smaller rock fragments fit into some of the larger voids. It may be noted that higher porosity does not necessarily imply higher permeability because the voids may not be interconnected. One good example of this is pumice stone which is a vesicular rock of volcanic origin in which the unconnected gas bubbles given rise to high porosity in solid state and making it light enough to float on water.

Figure 1

Well rounded coarse grained sediments have higher porosity than fine grained sediments because the grains do not fit together well

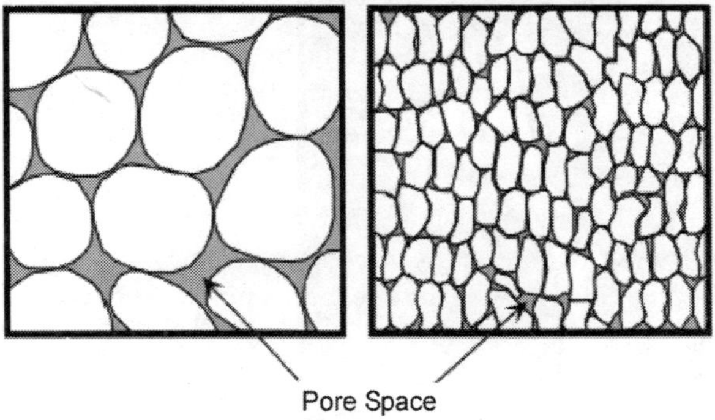

Source: Stephen A. Nelson, Tulane University.

A bed of soil or rock becomes permeable only when the holes and inter-facial spaces or passages are inter-connected to enable water to move between them. Such a porous bed is said to be permeable. Permeability is thus regarded as a measure of the degree to which the pore spaces are interconnected. The higher the permeability, the greater the ease with which water can pass through the rock (bed). Only a bed which is both porous and permeable qualifies to become an aquifer. It is obvious that truly solid or crystalline rocks, such as granite, or basalts cannot ordinarily host an aquifer.

Specific Yield and Specific Retention of Water of a Stratum

Ordinarily, if water is otherwise available, any stratum will hold water in the interstitial space between its particles, despite its natural tendency for downward infiltration under the gravitational force. This is because (a) forces of surface tension of water, capillarity and molecular attraction cause it to remain wrapped around the surfaces of the particles, and (b) where the inter-particle crevices are narrow enough for effective capillary action, water will always remain in them. Therefore, wherever the inter-particle spaces in the stratum are wide enough not to cause capillary retention water, water will move downwards, leaving behind that portion of it which had got stuck to the surface of the soil particles. This is shown schematically in Figure 2. It is this water that is the very source of soil moisture (green water in the terminology of water footprints). Therefore, when water is pumped

Figure 2

Picture showing a thin layer of water adhering to the surfaces of particles due to force of molecular attraction

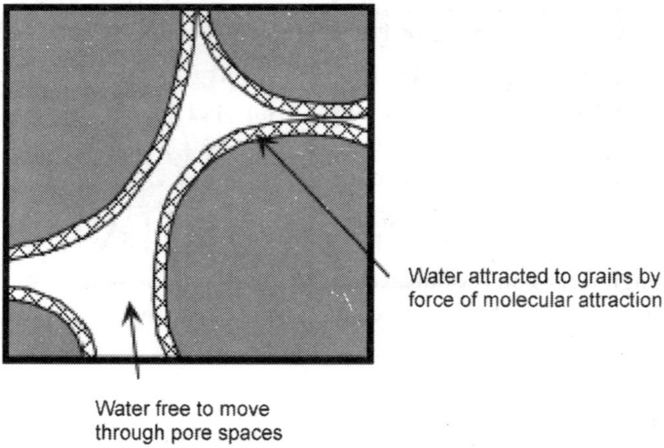

Source: Stephen A. Nelson, Tulane University.

out, all of it may not come out of the bed as some amount is retained by the bed. Hence the actual amount of water that can be extracted from a bed is called specific yield and that held back as specific retention.

Character of Ground Strata Suitable for Hosting Water

The process of infiltration of rain water to become ground water begins with the top surface layers rapidly absorbing rain water, before it runs off, and then letting it permeate into the underground strata leisurely. Thus the first obvious halt of water is in the unconsolidated top soil on earth, but the water holding capacity of every top soil is different. Its underneath may be of unconsolidated alluvium, granular materials, such as sand, gravel and clay and the consolidated (often called the bed rock) fine grained basaltic or granitic rocks, sandstone and limestone or any other metamorphosed rock like marble. Prima facie, any consolidated rock is not a good host for water. The point of importance is that the presence of porous surface plays the crucial role in the formation of sources of underground water. The unconsolidated material may consist of rock debris transported by glaciers or carried by turbulent streams and deposited wherever water velocity slowed down. It may consist of weathered and eroded particles of the bed rock that form a loose granular, alluvial soil or silt. They can store large quantities of water.

The consolidated rocks generally lie buried beneath unconsolidated bed/s or may outcrop at the land surface. Fossil aquifers are usually found in the consolidated rocks.

Ground Water: An Important Step in Hydrological Cycle

Underground water is regarded as a constituent part of the Earth's hydrological cycle which has been perpetually running with remarkable precision, since ages. The water cycles involves three basic steps: those of (i) evaporation from the surface of water bodies or transpiration from leaves, (ii) precipitation from the moist air as rain/snow/ice and (iii) transportation of the water back to the sea through rivers on the surface and through ground water aquifers. The cycle is shown schematically in Figure 3.

The paths and influences of water through earth's ecosystems are extremely complex and are not completely understood. A significant point is that in the hydrological cycle, all the three states of matter, i.e. solid (ice and snow), liquid and gaseous, vapour phase, as well as all its habitats are involved. The quantities staying in different types of habitats have remained statistically the same over ages.

Figure 3

The global water or hydrologic cycle pathways

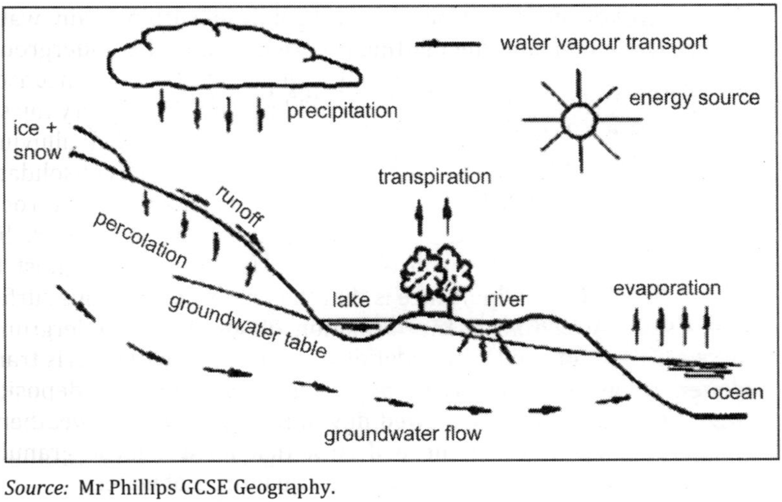

Source: Mr Phillips GCSE Geography.

Formation of Aquifers: Structural Nature of Ground Water Habitats

Only those underground strata that hold/store water in large enough quantities that can be tapped economically for utilization on the surface are called aquifers. On this definition, water of many dug wells is not regarded as coming from an aquifer; it is simply the water which has oozed out from the filled pores and inter-particle passages of unconsolidated porous land mass and has collected itself in the dug well.

Whilst the amount of water a rock can hold depends upon its porosity (size and amount of void spaces), its movement is determined by the permeability, i.e. ease with which water can move in the rock. Its downward movement ceases when it strikes an impervious stratum, called the bed rock. As the infiltration from the surface continues unabated, water starts building up its 'reservoir' on the impermeable stratum but in the permeable stratum. The downward movement could also be partially blocked by the following situations:

- When it strikes a rock/stratum, the permeability of which is less than that of the upper stratum, such a stratum offering partial resistance to the further downward movement of water is called an 'aquitard', giving rise to the formation of a pseudo-aquifer over itself, called a 'perched aquifer.'
- On its reaching very deep depths, where pressure from the huge load of the overburden causes consolidation of the aggregated particles through collapse of interstitial spaces, the stratum then behaves either as a true solid or as one with reduced permeability.
- The porosity of the aquifer bed could be greatly reduced if the percolating or infiltrating water also carried a proportion of finer material, such as silt or clay with it which block/clog the inter-particle passages.

Primary and Secondary Porosity of Strata

It may now be appreciated that the ability to hold water (and thus to act as storehouse for underground water) for all types of rocks and sediments would not be identical. Unconsolidated granular sediments, such as the alluvial plains formed by rivers, sand/gravel, contain pore space between adjoining grains, and their aggregated volume may be in excess of 30 per cent of the volume of the stratum. The porosity is called primary and the resulting aquifer defined as inter-granular aquifer. If such an aquifer forming stratum is spread over an area of 1 sq. km and is 1 m thick, it could hold 3,00,000 cu. m water. Now, assuming there is 50 mm rainfall in the area and all the water permeates downwards to the aquifer, the water level in the stratum would rise by a mere 170 mm.

It is, thus, clear that ground water aquifers cannot occur in highly consolidated rocks, such as igneous or metamorphic rocks with low permeability or where porosity is low. But such rocks may contain fractures or fissures, small cracks, pore spaces and spaces between layers in sedimentary rocks in which water could reside. Almost all rocks have them through geological and weathering activities, though to varying extents. The total volume of such fractures, etc. rarely exceeds 1 per cent of stratum volume. Further, the bed rocks may contain vertical fractures that may intersect other fractures, thus enabling water to move from one layer to another. Moving water over a long period of time over and through rocks (limestone and dolomite) containing water soluble minerals may form voids and channels. As a result, these fractures become enlarged and form fissures, caves and caverns (a cavern is a cave with many interconnected chambers). Water can reside in these 'hollows', in otherwise impervious rocks. As these voids are interconnected, water can also move along their slopes and travel long distances, through inter-connected spaces.

The aquifers formed therein are often called 'fractured aquifers', and the porosity is defined as secondary porosity. The two types of porosity are schematically shown in Figure 4.

Therefore, the rate of infiltration may be very slow, or nil, in certain types of rock strata such as marble and granite or fast in fractured and fissured sandstones and limestones or faster still in sandy soil.

Thus, an aquifer is water body that resides within a porous and permeable saturated rock through which it can also move. Besides, rock types as sandstone conglomerate, fractured limestone and unconsolidated gravel and sand, fractured volcanic rocks. In hydro-geological language, we may state that there are basically two classes of aquifers:

Figure 4

Schematic illustration of reduction of primary porosity and increased secondary porosity by solution

Primary porosity

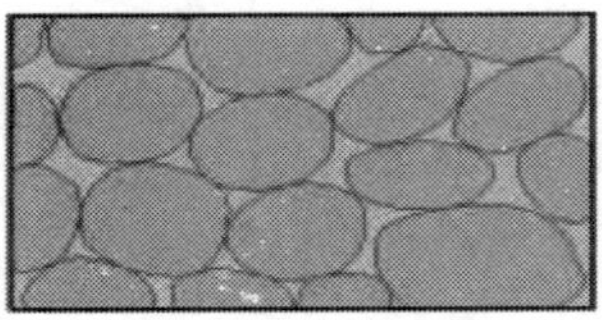

(a) Unconsolidated well-sorted sand; high porosity

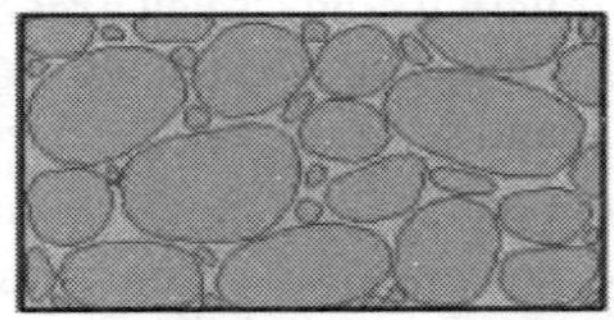

(b) Sand porosity reduced by admixture of fines or cementation

Secondary porosity

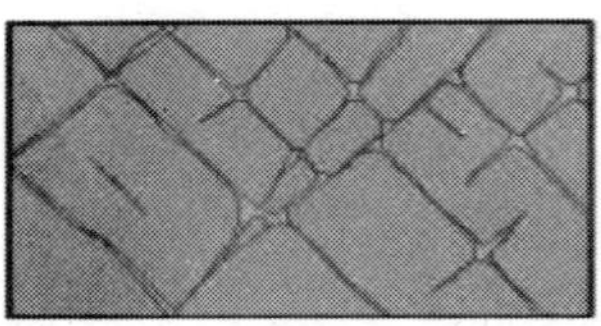

(c) Consolidated rock rendered porous by fracturing e.g. crystalline basement

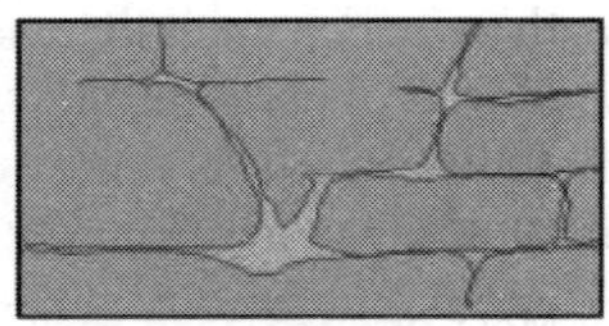

(d) Consolidated fractured rock with porosity increased by solution

Source: UNEP Natural Environment Research Council, 2003 Publication.

Alluvial Aquifers of the Sindhu-Ganga Plains

These aquifers have significant volume of spaces for storing or holding water within the aggregate of alluvial particles. These areas also cover the alluvial sediments of river basins, coastal and deltaic tracts. They are by far the most significant ground water reservoirs for large scale and extensive development. They are valuable source of ground water in northern India. Because of the ease with which water can be withdrawn from these aquifers, they are heavily exploited and face the risk of irreversible overexploitation.

Fissured Formations of Aquifers in Hard Consolidated Rocks of Peninsular and Central India

They are spread over almost two-third of the country and have negligible primary porosity; but are faulted and fissured due to intense geological activity and weathering. From hydrogeological point of view, fissured rock formations are broadly classified into four types:

- Igneous and metamorphic rocks: Ground water yield is variable and depends on rock type and extent of fracturing and fissuring.
- Volcanic rocks: The predominant rocks are the basaltic lava flows of Deccan Traps, made up through the geological consolidation of several plateaus. The Deccan Traps have usually poor to moderate permeabilities depending on the presence of primary and secondary pore spaces and fractures.
- Consolidated sedimentary rock formations: These consist of conglomerates, sandstones, shales, slates and quartzites which occur in the Cuddapah and Vindhyan ranges. The extent of the presence of bedding planes, joints, contact zones and fractures controls the ground water yield of aquifers.
- Carbonate rocks: Limestones in the Cuddapah, Vindhyan and Bijawar groups of rocks dominates the carbonate rocks other than marbles and dolomites. In carbonate rocks, the circulation of water creates solution cavities thereby increasing the permeability of the aquifers. Solution activity leads to widely contrasting permeability within short distances in such rocks.

Most of such hard rock formations are found in central peninsular India where they are typically overlaid by soil. They have poor permeability which makes the recharge of their aquifers through rains a slow process. Therefore, the porosity and permeability of the soil cover over them is important. The soils act as a transit store for rain water, and hold it to infiltrate into the fractured aquifers.

Aquifer water may also naturally discharge on surface of the Earth as springs at appropriate geological formations or directly seep into a river when fissured rock outcrops itself in the river valley; or even in sea.

It is interesting to note that in the year 2013, very large aquifers of fresh water or water of low salinity were discovered under continental shelves of Australia, China, North America and South Africa with total water volume estimated at half a million cu. km. Geological experts state that these deposits were created thousands of years ago when ocean levels were lower and rain water made its way into the ground in land areas, i.e. in those that were not submerged, until the last ice age ended over 20,000 years ago and submerged them.

Nature of Sub-Surface Water Distribution and Water Table or the Saturated Water Zone

This may be appreciated with reference to Figure 5. As the rain water percolates down into the soil, it may do so selectively through some of the available inter-particles passages, leaving the others filled with air at atmospheric pressure. This is the zone of aeration or the so-called 'vadose' or unsaturated zone. The unsaturated material above the water table contains both air and water in the spaces between the rock particles. It ordinarily consists of three zones: the belt of moist soil; the intermediate or the vadose zone and the capillary fringe.

The belt of moist soil contains water held by molecular attraction. It lies just below the land surface and extends downwards; this zone harbours and supports the root system of vegetation on the surface as the interstices are aerated as well. It is important to note that this moisture in the soil is not available to seep down to the wells but is of great importance to horticulture and agriculture.

This is followed by the intermediate vadose zone. In this zone, the amount of water exceeds that which could be held by molecular attraction alone. It is this water that takes part in the process of downward infiltration.

When further downward seepage is arrested by an impervious stratum, water begins to build up, progressively occupying all the vacant pores and inter-particle passages; thus constituting the so-called saturated zone. The maximum level above the impervious stratum to which such a saturated zone occurs is called the 'water table'. In the saturated zone below the water table, ground water fills in the spaces between unconsolidated particles. It may be noted that only alluvial or weathered soils can show water table.

The water table is located with reference to the surface by stating how far below the surface it occurs. Figure 5 also shows a rather narrow zone in which all the pores and inter-particle passages are filled with water due to capillary action.

In terms of hydrology, water table is defined as the surface where the pressure head is almost equal to atmospheric pressure, (i.e. the gauge pressure is nil). If a well is dug in the ground deeper than the water table, water would collect in the well up to this level; capillary fringe zone would not affect it because capillary action does not happen in large diameter bores. One can see this phenomenon happen by filling an earthen pot with water; the moist line on the pot wall is higher than the actual level of water in the earthen pot. For the same reason, capillary fringe zone would not be found over saturated bed made mostly of gravel and of such other coarse particles.

The capillary rise in alluvial soils could vary in the range 30–180 cm, depending on the fineness of the particles of the alluvium. The finer the soil, the thicker the capillary fringe. The capillary fringe may be very thin in coarse-grained sediments, in which the capillary action is negligible. The water in the capillary fringe is not available to wells, which must be deepened to the zone of saturation before water will enter them.

Figure 5

Schematically water table and the saturated and unsaturated zones

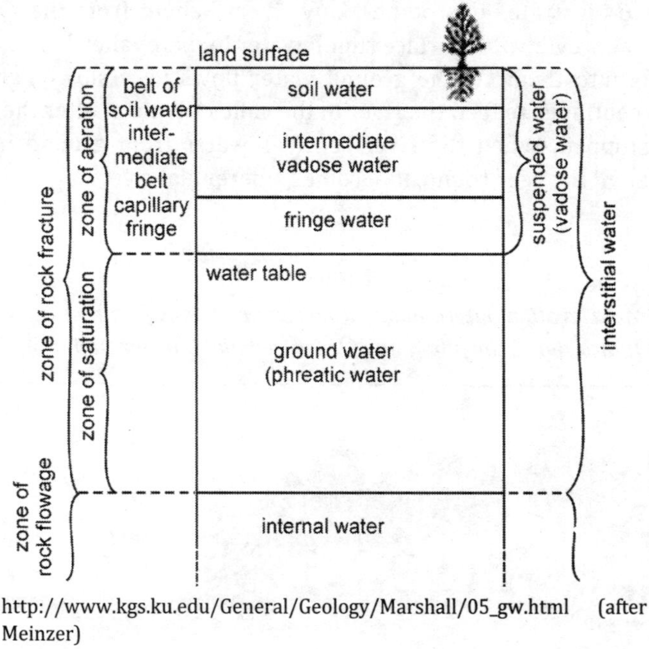

Source: http://www.kgs.ku.edu/General/Geology/Marshall/05_gw.html (after OE Meinzer)

It may be appreciated that the zone of capillary fringe exists because of saturated ground water table. The capillary fringe zone occurs within the vadose zone (i.e. the unsaturated but aerated zone), and homes the root system of the trees above the ground. Since it sucks water from the saturated zone and pushes it upwards into the zone supporting the root system of vegetation on the surface, it acts as the Nature's water sprinkler irrigation system to provide water just where it is needed. It is because of this phenomenon that the soil always maintains its moisture, of which the ultimate beneficiary is mankind. It may be appreciated that the location of capillary fringe is directly related to the water table. In the event of water table going down, this zone will also go down. A situation may develop that it goes so down, beyond the reach of the roots of trees above; the soil above dries up and so do the trees.

Ground Water Feeds Rivers

The water table also slopes with the sloping of the bed rock. When the slope leads to a river, the water table may outcrop in the river. In the event of this happening, the ground water feeds the river as well. Figure 6 illustrates this point. Many rivers benefit from the ground water rains even after surface runoff water in their valley has stopped flowing into them. As the ground water flows very slowly, ground water continues to feed the river in the valley till much after the rains have stopped. When the river receives water from ground in this manner all the year round, it becomes a perennial river.

Figure 6

Saturated stratum outcropping into a river bed, thus feeding the river as well. It also shows the filling up of fractured and inter-granular aquifers.

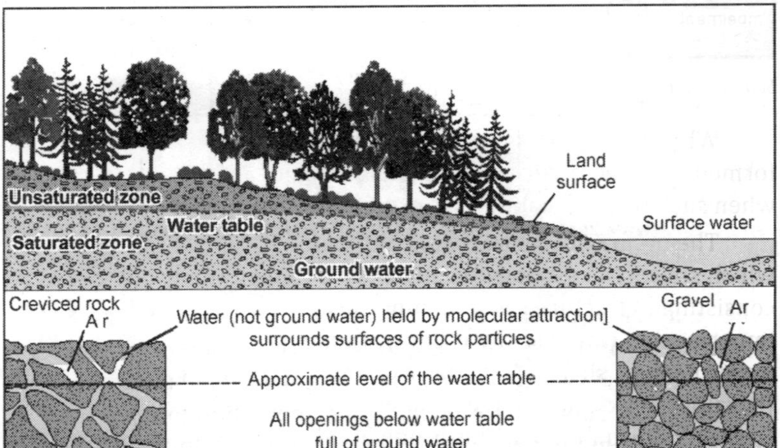

Source: US Geological Survey.

Ground Water Appearing as Springs on Surface: Geological Faults

In the foregoing discussion, we have stated at a few places that underground water manifests itself at few geologically crafted places in the form of water spurting out of springs. Scientifically, spring can appear in area on the surface of the earth where the impermeable rock/stratum (called an aquiclude) supporting the aquifer stratum is exposed to atmosphere enabling water from the stratum flows out. Clearly, such a situation of strata splitting into relatively displaced blocks, with a face exposed to atmosphere, can occur only through geological activity. The plane of split or parting is called a 'fault'. The resulting situation is shown in Figure 7. Whether water will merely seep out or form a well defined spring depends on the flow of water in the bed above; when abundant, water will gush out as spring.

Figure 7

Geological situations favouring appearance of natural springs on the surface

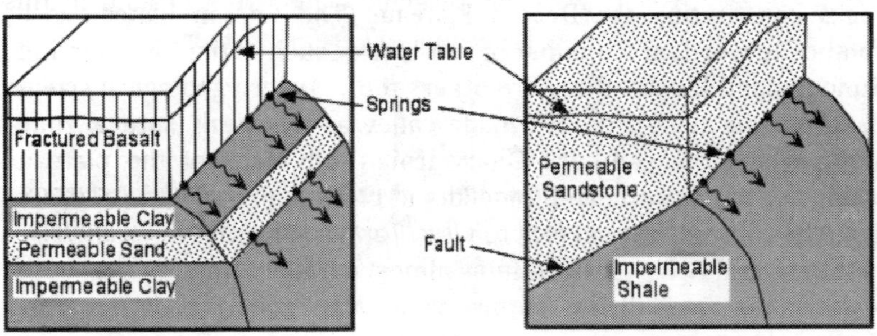

Source: Stephen A. Nelson, Tulane University.

When the splitting extends over long distances, the valley so formed is called a rift valley. Water oozes or gushes out as springs when such rock formations are exposed on the surface.

The occurrence of the rift is illustrated with the help of Figure 7 which shows the sections through two assumed types of strata: one, consisting of fractured basalt, impermeable clay, permeable clay and impermeable clay and the other of permeable sandstone resting on impermeable shale. A good example where fractured igneous (basaltic) rocks act as a storehouse of rain water comes from the Deccan Plateau in India. This is one of the largest volcanic formations in the world, formed by massive flows of lava (molten rock) from erupting volcanoes; it is made of igneous rocks, with mineralogical name 'basalts'. Basalt is a fine gained rock and is most widespread of all igneous rocks and comprises more than 90 per cent of all volcanic rocks and is the most abundant bed rock of the earth. Experts have dated the multiple volcanic eruptions to have occurred 67 million years ago. Though by itself impervious (therefore incapable of hosting water), the rocks of the Deccan Plateau had to bear the brunt of massive geologic or tectonic activities as a result of which the impermeable rock mass got heavily faulted and fissured. Since the faults and fissures are mutually interconnected, they are not only a good host for rain water but also provide channels for the water to move or flow in the direction of slope of the strata.

Rivers Narmada and Damodar: Typical Examples of Beneficiaries of Rain Water Stored in Rocky Strata Faults

Narmada river, one of the most sacred rivers of India, flows east to west in such a valley and the beneficiary of rain water stored in its underground catchment area in form of the fractures and fissures in rocks constituting the Deccan Plateau. (The Deccan Plateau is a composite word for a number of sub-blocks, such as the Dharwar and Bundelkhand Blocks, there are others too). Modern geological scientists in India regard the Narmada valley as the weak 'suture' zone along which the southern Dharwarnian sub-block of the plateau welded with its northern Bundelkhand block many millions of years ago. The suture is spread east to west. Tapti (or Tapi) river is another east to west flowing river, running almost parallel to, and to the south of Narmada river in the 'suture' zone. Narmada river is not only perennial but sustains several dams, besides the giant Sardar Sarovar, near Ahmedabad and is expected to play an important role in the Kalpsar Project which seeks to create a reservoir of fresh water in the sea. The Narmada waters have brought about quantum jumps in the agricultural and horticultural produce of Madhya Pradesh, and Gujarat, besides meeting domestic and industrial demands. The waters of Tapti river are also heavily harvested through dams and weirs, including the Ukai Dam in Gujarat.

Another important river of India – Damodar river – is also a beneficiary of water in the cracks and fissures of igneous rocks. It originates at a place called Chandwa in Jharkhand State in the Chota Nagpur sub-plateau of the Deccan Plateau. Geologists regard the Chota Nagpur plateau as an extensive area of land which was thrust above the general land due to collision of the tectonic plates, some 50 million years ago. It was the first interface of the Deccan with massive Eurasian continent which obstructed further movement. The region saw violent geological activity as a consequence of which the Chota Nagpur Plateau was uplifted. The rocky strata of the plateau, therefore, got heavily fragmented and highly dissected, resulting in the formation of many precipitous escarpments. There is a variety of rock formations. The metamorphosed granitic rocks are associated with porous lateritic rocks with sand and gravel which are good hosts of water. The area is drained by several streams and rivers which flow in all directions in highly faulted valleys. Damodar river originates and flows in a rift valley and receives the waters of many tributaries. This means that tributaries of Damodar River drain the waters of the monsoon rains of the entire sub-plateau into the river within a short time. As a consequence of which Damodar river swells in the rainy season and causes devastating floods, as if in rage and fury. The river was known as the 'sorrow of Bengal', until tamed by several dams that were built by the Damodar Valley Corporation in the post-1950 period. There are many waterfalls where rivers coming from the uplands of the plateau descend precipitously and enter areas of significantly lower heights.

Thus, in so far as hosting water is concerned, the presence of faults in the rocks is of primary importance.

Types of Aquifers: Confined and Unconfined Aquifers

In the foregoing discussion, we have classified the aquifers on the basis of the structural constitution of their body mass. Another way to classify them is on the basis of the nature of their roof and bed rocks. Of course, in any method of classification, the bed rock must necessarily be impervious. When the roof stratum of the aquifer itself is pervious, it is said to be an unconfined aquifer. Aquifers of this type are more common. They have one layer of impermeable material beneath the saturated zone. They are relatively easy to recharge with percolating rain water. It may also be specifically stated that unconfined aquifers exist below the water table only. They are sometimes referred as water table or phreatic aquifers because the upper boundary of their existence is restricted by the water table. The term 'phreatic surface' indicates the location where the pore water pressure is equal to the atmospheric pressure, i.e. the pressure head is zero. This surface normally coincides with the water table where the rock and soil is saturated with water. The phreatic surface slopes in the direction of movement of water in an unconfined aquifer. Ground water is always moving by the force of gravity from recharge areas to discharge areas. Ground water movement in most areas is slow – a few metres per year. But, in more permeable zones, such as solution channels in limestone, movement can be as much as several metres per day.

When the roof rock too is impervious, the aquifer which is sandwiched between the two impermeable rock formations is called 'confined aquifer'. As it cannot receive water infiltrating from the top, its only source of water input is through the outcrop of its rock at the surface, i.e. where it peeps out at the surface. However, such a rigid situation does not usually occur in nature as almost all the rocky strata are always heavily faulted and contain many cracks through which water from the upper permeable stratum trickles down to bank in the confined aquifer. Confined aquifers are rarer than unconfined aquifers, and occur when the saturated zone lies between two layers of impermeable rock. These types of aquifers result in wells that do

Figure 8

Structural dispensation of impermeable and permeable stratum for the formation of confined and unconfined aquifers.

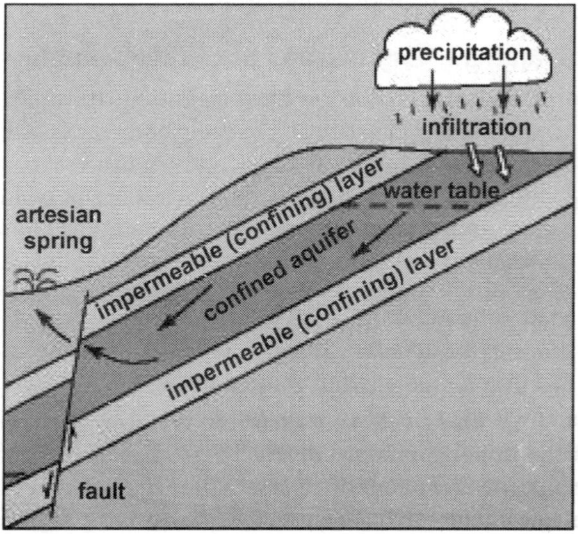

Source: Kansas Geological Survery, USA.

not require pumping, or artesian wells. The geological structure required for the occurrence of two types of aquifers is shown in Figure 8. The Figure also shows the formation of a geological fault. In the situation shown in Figure 8, a confined aquifer is feeding the bed ground of the unconfined aquifer. In the event of the valley hosting a river, a confined aquifer may feed the river as well specially if the river flows in the fault zone. Many rivers flow in the fault zones or suture zones.

Figure 9 shows the conditions for the occurrence of Artesian well. In this context, an Artesian well may be regarded as a special kind of confined aquifer.

Figure 9

Outcrops of sedimentary rocks that have formed confined aquifer; also showing the occurrence of artesian well and artesian spring

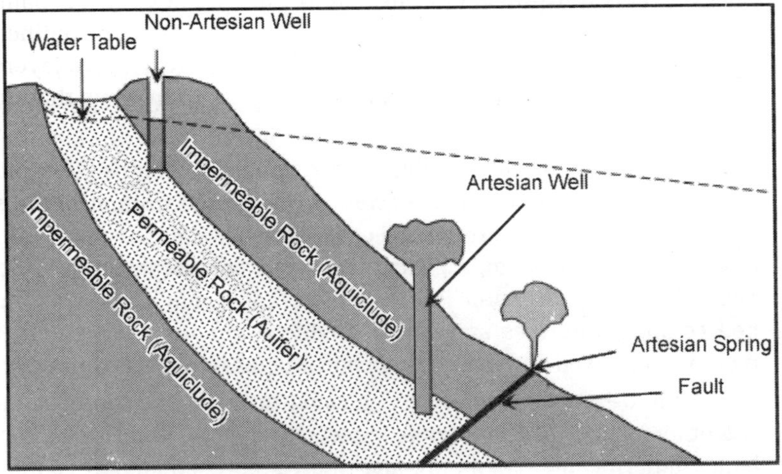

Source: Stephen A. Nelson, Tulane University.

In confined aquifers, however, the pore water pressure could be several times more than the atmospheric. If a well is dug vertically to reach a bed of confined aquifer, water will rise effortlessly in the well, which is now called Artesian well, shown in Figure 9. But it must lie below the phreatic surface (shown by dotted line in Figure 9) to become 'Artesian'; if not, it is simply a non-Artesian well. In the event of a geological fault having occurred in the strata, an Artesian spring may appear suitably.

It may be stated that the permeation of water through porous rocks/faults is an extremely slow process, unlike that in alluvial soil. Since there is no possibility of the withdrawn water being replaced by fresh inputs, any withdrawal of water reduces the hydrostatic pressure in the pores. In course of time, the well may lose its Artesian character. The US Geological survey has pointed out that 60 years ago, Artesian aquifers were fairly common in the US but today they have become rare because of large withdrawals of water from them.

Deep underground aquifers are as a rule of the confined category and it is practically impossible to recharge them. According to hydro-geologists, the residence time of water in them may vary between thousands of years to millions. It is reported that the water in the emptying or already emptied aquifers in Saudi Arabia is/was at least 20,000 years old.

Multi-Layered Aquifer System

Although any porous stratum can hold only one aquifer, a strata of sedimentary rocks may host more than one aquifer system because of differential permeability and geological characteristics of individual stratum the strata, including disruptions in the form of faults and cracks. If the stratum contains water soluble minerals, they get partially dissolved in the aquifer water. Thus, when water percolates through deposits of limestone and chalk, which are largely made up of calcium and magnesium, it picks them up as bicarbonates, chlorides and sulphates; such water is called 'hard water'. Similarly when the bed contains fluorides and water soluble compounds of arsenic, it picks them as well and makes them harmful for human consumption.

A multi-layered aquifer system constituted by one confined and the other unconfined aquifer is shown in Figure 10 with bed rock being impermeable. The figure also show a perched aquifer saddled on a block of impermeable stratum but below the general in the vadose zone water table. More commonly, however, perched aquifers are saddled on blocks of impervious rocks in the vadose zone above the water table. The dimensions of a perched aquifer are usually small and the volume of water they contain is sensitive to climatic conditions, principally rain.

Stratification in sedimentary rocks is a naturally expected occurrence but under certain conditions, sufficient stratification could be found in alluvial beds as well. This may result in the bed showing differential permeability along a vertical section.

Figure 10

A multi-layered aquifer system with perched aquifer water table, self-flowing artesian well, an artesian well and a water table well to illustrate the differences between them

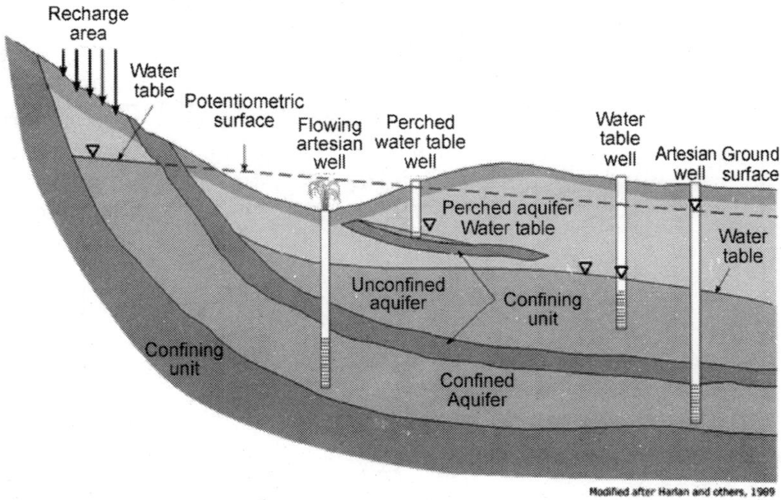

Source: Colorado Geological Survey, USA.

Effect of Pumping Out Large Quantities of Water from Unconfined Aquifers through Deep Bore Well

As ground water is slowly pumped out through a number of scattered pumps of small capacity, the water table uniformly falls over a wide area until replenished. This explains why the water table invariably falls down in summer but rises up again after the rains. The variations in water table depth are significantly noticeable in hot tropical climatic zones where there are marked dry and wet seasons. Where the ground water is withdrawn in large quantities through a single bore well with the help of a powerful pump, the resulting situation could assume grave dimensions as a cone of water table depression is thereby created around the bore pump, as shown in Figure 11. It is simply not an innocuous phenomenon as it exerts far reaching consequences. The drop is more in the immediate vicinity of the pump but rises up very gradually away from it. Effectively, it means that if a very deep bore well is functioning with powerful pump, all bore wells, less deep than the functioning well, over a wide area could dry up, without

Figure 11

Diagrammatic section showing drawdown, cone of depression, and radius of influence around a pumping well

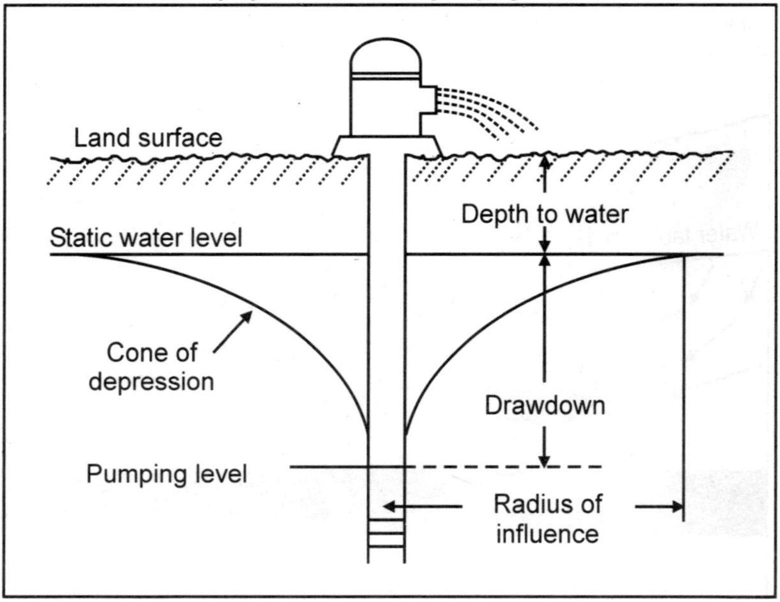

Source: Kansas Geological Survey, USA.

any one ever knowing who the real culprit is. In other words, the huge centralized withdrawal of ground water has left little or no fresh water for others who also depend upon it. Even though some sand and gravel aquifers are very long and thin, they are recharged locally by precipitation. While the deposits can extend for miles, the usage of ground water only affects the water table locally.

The resulting situation is pregnant with ecological disaster. An attempt is made to explain this situation with the help of Figure 12. The situation is particularly bad if there is a flowing river/stream/ravine nearby. If initially, the water table was above the river bed, ground water reinforced water flow in the river. When the water table falls below the river bed with excessive pumping out of the ground water, the situation is reversed as the river itself begins to feed the bore well, leaving little or no water for its downstream path. Therefore, installation of deep and powerful bore well pumps in drought prone areas should be regulated and not left to individuals.

Figure 12

The figure at the top shows the water table and the natural flow of aquifer water to the river. The bottom figure shows the effect of heavy pumping out lowering the water table enough to cause the river to feed the ground water

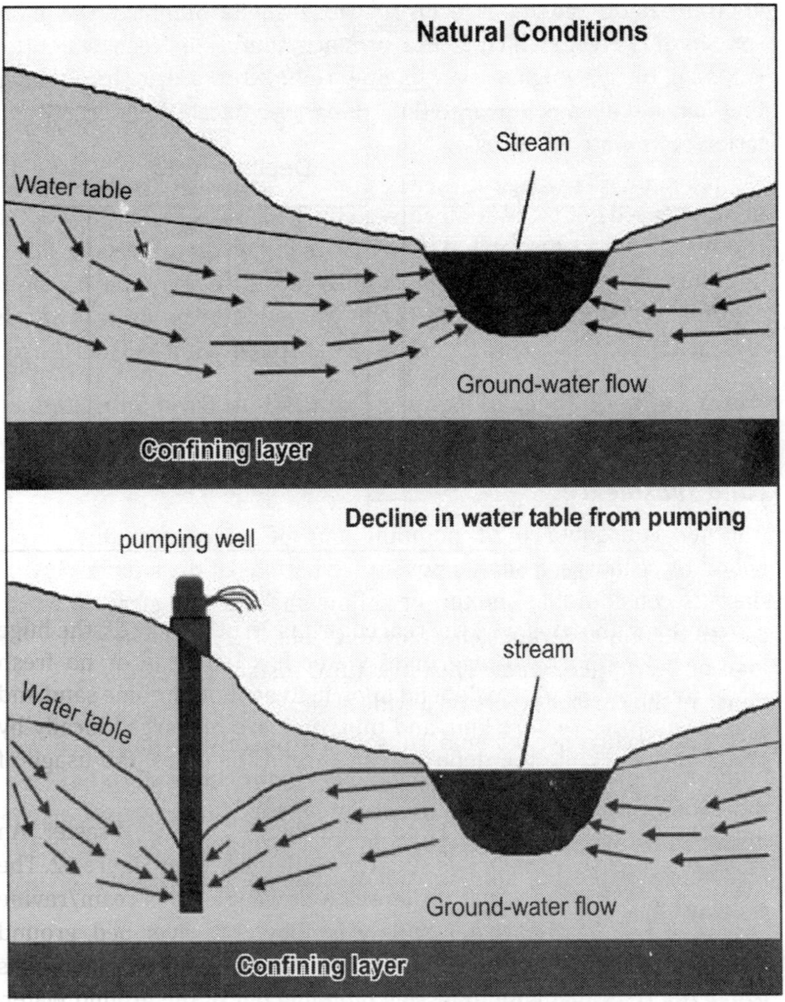

It has been explained in a preceding section that excessive withdrawal of water through deep bore wells can dry up a river. This phenomenon has been found to occur and many prominent rivers of the world are victims of it. Ground water depletions have led to the inability of the Yellow river in China to reach the ocean for months at a time. Its longest dry-spell is reported to have been 200 consecutive days, during which time the river did not flow into the ocean. The Colorado river in the US has many dams and diversion along its 2,333 km route to the sea that it rarely reaches its delta. Similarly, the Indus (or Sindhu) river, which is the primary source of fresh water for Pakistan, once a mighty river, is now reduced to a dribbling stream that may not always flow into the Arabian Sea because of over exploitation of its water in its basin.

In simple language it means that the amount of ground water being pumped out is not compensated by the natural recharge with the result that the water table had been continually going deeper down until it reached the rock bottom. In India too, the water table has been continually going down at many places. The situation in certain districts of western Karnataka, and Vidarbha in Maharashtra is particularly bad.

Other Consequences of Aquifer Exhaustion: Land Subsidence and Ground Water Pollution

Land Subsidence

Another consequence of pumping out of water, without corresponding recharge, from the porous formations of the aquifers is that there is considerable amount of hollow space in the stratum which causes the upper strata to become unstable. They collapse under the load of the upper strata. This structural instability results in subsidence of the surface layers on a wide scale.

The occurrence of land subsidence on this account is a well monitored and documented phenomenon in the USA. Figure 13 shows the 28 feet road subsidence observed by the US Geological Survey during time interval 1925–1977 in San Joaquim Valley California. Elsewhere in the US, land subsidence damaged roads, bridges and railway lines.

Geogenic Pollution

When the strata contain multiple aquifers, not necessarily all of fresh water, their waters remain within their boundaries in a state of hydrostatic equilibrium. But when fresh water is excessively withdrawn from one of them, the equilibrium is lost and water from others, yet under pressure, infiltrates into the one which is getting depleted, causing it to get polluted. This is called geogenic or natural pollution

Figure 13

Visible evidence of land subsidence recorded by US Geological Survey by indicating the levels of road in the 1925–77 period in San Joaquim Valley, California.

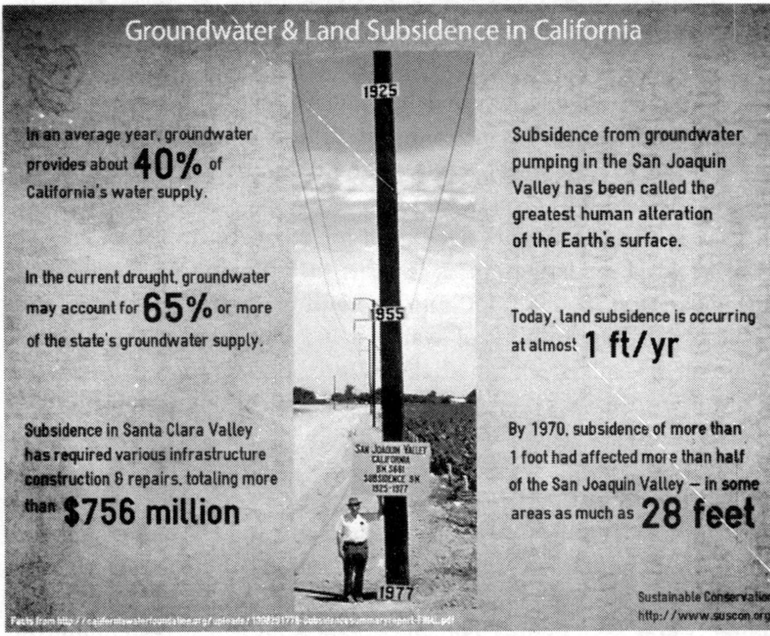

as it occurs as a result of hydro-geological processes that occur within the neighbouring strata. In this way, many fresh ground water aquifers in India are getting polluted with infiltration of water containing harmful compounds of arsenic, fluorides, nitrates or iron, i.e. their contents are in excess of the limits prescribed for drinking water. According to Central Ground Water Board estimates in 2015, 68 districts in the states of Haryana, Punjab, Uttar Pradesh, Bihar, Jharkhand, Chhattisgarh, West Bengal, Assam, Manipur and Karnataka have high arsenic contamination; 276 districts in 20 states with fluoride contamination; 387 districts in 21 states with nitrate contamination and 297 districts in 24 states with iron contamination. Geogenic pollution of underground water resources is a fairly widespread phenomenon in India due to unsustainable withdrawals of underground water.

In addition, underground water resources are also getting polluted due to:

- anthropogenic activities – overuse of fertilizers and pesticides in agricultural practices, industrial toxic effluents containing heavy metals like mercury, lead and radioactive waste pollution and pollution through landfills.
- Biogenic contaminants include pathogenic bacteria through open air faecal discharges, discharge of untreated sewage water, etc. and,
- In coastal regions, large withdrawals of fresh water provide opportunities to the sea to ingress into and occupy space in the emptying fresh water aquifers and, thus, turn the fresh water resource saline.

Saline water can neither be used for domestic applications, nor can it be used for irrigation as it turns toxic for the plants above a concentration of about 1,300 mg/litre and is not potable above 250 mg/litre. Further, salinity of water helps building up the sodium content in the soil, which causes the soils to be compact and hard, with reduced porosity, as a consequence of which the water holding capacity of the soil is considerably reduced.

Rains and Ground Water

It is readily acceptable that nearly all surface water sources would dry up on continued total failure of monsoon rains. The close relationship existing between rainfall and water table aquifers (unconfined aquifers) may be a hydro-geological reality but is less widely appreciated amongst common people. In the event of continued failure of the Indian monsoon, all resources of fresh water, both on surface and underground would dry up, as also agricultural farms using water from surface canals or underground water through dug and/or bore wells. Water is required in every nook and corner of the country, but there are wide dissimilarities in the distribution of rainfall across the country, as shown in Figure 14. If there are rainfall deficient areas like that of Thar desert in western Rajasthan or the Ladakh plateau, there are rainfall abundant regions of coastal regions on the west of the Western Ghats and in Assam and north-east states. Yet, regarding India as a unit, the country receives an average rainfall of 1,170 mm/year. On this basis, the country receives 4,000 billion cu. m (BCM) annually of fresh water through rains. Out of this volume of precipitation, (according to the estimates of Central Ground Water Board 2006/7), only 1869 BCM goes as average annual potential flow in rivers and streams. Only 1123 BCM was then assessed as the annual

Figure 14

Rainfall map of India

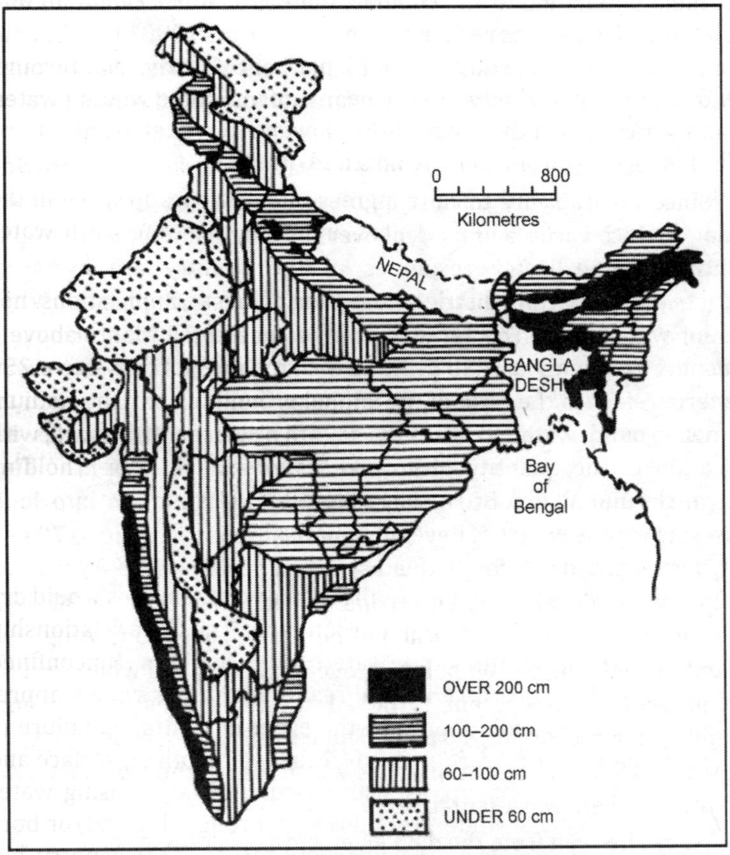

Source: Food and Agricultural Organisation of the United Nations.

utilizable water. This is due to constraints of topography and uneven distribution of rain water in various river basins which make it difficult to extract the entire 1869 BCM/year; out of this – 433 BCM/year and 690 BCM/year recharge the ground and surface water resources respectively. The overall contribution of rainfall to country's water resource is 68 per cent and the rest 32 per cent comes from other resources, such as canal seepage, return flow from agriculture, recharge from natural ponds/lakes and manmade reservoirs. As of then, total water use, including ground water was assessed at 634 BCM, of which 83 per cent was for irrigation.

The demand for water for 2010 was then projected to grow to 813 BCM, to 1093 BCM by 2025 and to 1447 BCM by 2050. Due to increasing population in the country, the national per capita annual availability of water has reduced from 1816 cu. m in 2001 to 1,544 cu. m in 2011. This is a reduction of 15 per cent. Clearly, the demand would outstrip availability in very near future. People would turn to ground water to bail them out of the shortages through deeper bore wells, taking a cue from Saudi Arabia and others.

Since the capacity to host aquifers depends on local geomorphology, which varies a great deal over the length and breadth of the country, ground water resources in the country are assessed district-wise and within districts according to blocks, *mandals, talukas* or mini watersheds. The term ground water exploitation or development is frequently used in defining the ground water environment. The term refers to the ratio of the annual ground water extraction to the net ground water availability. In other words, it provides an indication of the quantity of ground water available for use. The Central Ground Water Board has classified the situation into four classes: when the extent of development in any district is up to 70 per cent, it is regarded safe for further use, when between 70–90 per cent, cautious water development is recommended, when between 90–100 per cent, the situation is critical and need intensive monitoring and evaluation and when >100 per cent, overexploited, and where future ground water development is linked to water conservation and ground water recharging systems. The current assessment methodology uses 15,640 observation wells which makes the available data indicative and not representative.

As can be seen from the data given in the preceding sections, the availability of surface waters is greater than ground water in India. However, owing to the decentralized availability of ground water, it is easily accessible and takes the largest share in irrigation and water supply. Over the years, there has been a decrease in the use of surface water through canals and a continuous increase in ground water utilization for irrigation.

Ground Water Availability in Indian States

There are wide differences amongst the different Indian states in respect of availability of water. Statewise data for water resources up to a depth of 600 m is shown in Table 1.

Table 1

State-wise ground water availability and utilization through active aquifers (up to 600 m below mean sea level)

State	*Net Annual Ground Water Availability BCM*	*Annual Ground Water Draft* BCM*	*Projected Demand for Domestic and Industry Uses Upto 2025 BCM*	*Ground Water Availability for Future Irrigation BCM*
Andhra Pradesh	32.95	14.90	2.67	17.65
Arunachal Pradesh	2.30	0.0008	0.009	2.29
Assam	24.89	5.44	0.98	19.06
Bihar	27.42	10.77	2.14	15.89
Chhattisgarh	13.68	2.80	0.70	10.67
Delhi	0.28	0.48	0.57	0.00
Goa	0.27	0.07	0.04	0.18
Gujarat	15.02	11.49	1.48	3.05
Haryana	8.63	9.45	0.60	–1.07
Himachal Pradesh	0.39	0.12	0.04	0.25
Jammu Kashmir	2.43	0.33	0.42	1.92
Jharkhand	5.25	1.09	0.56	3.99
Karnataka	15.30	10.71	1.41	6.48
Kerala	6.23	2.92	1.40	3.07
Madhya Pradesh	35.33	17.12	1,74	17.51
Maharashtra	31.21	15.09	1.52	16.10
Manipur	0.34	0.002	0.02	0.31
Meghalaya	1.04	0.002	0.10	0.94
Mizoram	0.04	0.0004	0.0008	0.04
Nagaland	0.32	0.009	0.03	0.30
Odisha	21.01	3.85	1.22	16.78
Punjab	21.44	31.16	1.00	–9.89
Rajasthan	10.38	12.99	2.72	–3.94
Sikkim	0.08	0.01	0.02	0.05
Tamil Nadu	20.76	17.65	0.91	3.08
Tripura	1.97	0.17	0.20	1.69
Uttar Pradesh	70.18	48.78	5.30	19.52
Uttaranchal	2.10	1.39	0.08	0.68
W Bengal	27.46	11.65	1.24	15.32
Total All State	398.70	230.44	29.12	161.92
Total All India**	399.25	230.62	29.16	162.20

* Includes requirements for irrigation, domestic and industrial. ** includes Union Territories.

Source: Central Ground Water Board, 2006.

The data camouflages some harsh realities; for example it shows that some states, like Andhra Pradesh, Karnataka and Maharashtra, are in apparent happy situation of having water available for utilization but the grim truth is that within the states, particularly in Maharashtra and Karnataka, ground water resources have completely dried up in certain areas. Anomalous situation of this kind is built up because at state level estimates, regions of heaviest rainfall along the western side of Western Ghats are also taken into consideration. As of now, these waters are wasted into the Arabian Sea and there is no possibility of their utilization for the benefit of the state in foreseeable future. Close study of the data reveals another unsavoury truth that a few states like Rajasthan, Punjab, Haryana and Delhi are pumping out more water than is being charged. During the last decade and a half, western Uttar Pradesh has also been drawing out more ground water than being charged into it as evidenced by the continuing falling water table.

Nature had not intended to make the distribution of surface and ground water resources subservient to political divisions and to this extent any discussion of the resources in terms of state boundaries strikes discordant notes with Nature. It is, perhaps, more appropriate to divide the country either as, in terms of geomorphologic features, ground water habitats or river basins. A short discussion is presented on the latter as it has been used to plan and develop inter-basin connections to harmonize distribution of water across the country as understanding of the former necessitates deeper understanding of the geological sciences.

Potential for Ground Water Recharge

India has four major types of geomorphic forms, each with a different kind of ground water scenario. These are summarized in Table 2.

Table 2

Potential for ground water charging according to regional geomorphology

Geomorphic System	*Coverage*	*Potential for Recharge*
Alluvial	Sindhu-Ganga-Brahmaputra Plains	Enormous quantities up to 600 m. High rainfall; hence recharge is assured. Can support large scale development through deep bore wells.
	Coastal Plains	Reasonably extensive aquifers but risk of saline water intrusion.

Cont'd...

...Cont'd

	Desert Plains of Rajasthan and Gujarat	Scanty rainfall, no recharge potential. Available at great depths. Salinity hazard.
Consolidated/semi-consolidated sedimentary rocks, basalts and crystalline rocks	Peninsular India	Availability depends upon secondary porosity developed due to weathering and fracturing; therefore varying yields. Available at varying depths between 20–40 m in some areas and deeper depths 100–200 m in others.
Hilly		Low storage capacity due to quick runoff.

Source: Central Ground Water Board.

Table 3

Passive fresh ground water resource

State	*Alluvium/Unconsolidated Rocks Cu. km*	*Hard Rocks Cu. km*	*Total, cu. km*
Andhra Pradesh	76	26	102
Assam	920	0	920
Bihar	2,557	11	2,568
Gujarat	92	12	104
Haryana	420	1	421
Himachal Pradesh	13	0	13
Jammu Kashmir	35	0	35
Karnataka	00	17	17
Kerala	05	6	11
Madhya Pradesh	14	27	41
Maharashtra	16	22	38
Odisha	162	13	175
Punjab	910	00	910
Rajasthan	115	13	128
Tamil Nadu	98	00	98
Tripura	101	0	101
Uttar Pradesh	3,470	30	3,500
West Bengal	1,625	01	1,626
Delhi and Chandigarh	4	0	0
Total	10,633	179	10,812

Source: Ministry of Water Resources, Government of India 1999.

Passive or Inactive Ground Water (Aquifers)

Ground water experts are unanimous in stating that Indian land mass holds very large quantities of hitherto untapped ground water below a depth of 600 m and their estimate is provided in Table 3. These deep aquifers are called 'passive' or 'inactive' because, unlike those above them, they are not chargeable with rain water and are not affected by fluctuations in the water table. Geologists tell us that water in these aquifers had accumulated many thousand years ago and is not renewable. They are many times referred to as fossil aquifers.

Knowledge of fundamentals of hydro-dynamics tells us that the hydrostatic pressure deeper down below the water table increases with the depth but at any point of time there is dynamic equilibrium in the bed, i.e. deep down water under higher pressure does not disturb the water in the upper layers which otherwise exerts lower hydrostatic pressure. It has been stated earlier that the percolating water picks up the soluble minerals from the rock beds. If the bed had water soluble compounds of arsenic or fluorides, they were dissolved away by the water percolating through them several thousand years ago, almost as first-come-first-served basis. In other words, water in the so-called passive aquifers is rich in dissolved chemicals. Obviously, the water that passed subsequently through the same bed and now resting in the upper active aquifers is relatively purer.

India: Top Worst Performing Nations in Safe Drinking Water Access

In terms of number of people, India has the largest number of people without access to safe drinking water, according to a report released by WaterAid (WaterAid is an international organization whose mission is to transform the lives of the poorest and most marginalized people by improving access to safe water, sanitation and hygiene) on World Water Day, 2016. The global data is as follows:

India	—	7,57,77,997 (which is about 5% of the population)
China	—	6,31,66,533
Nigeria	—	5,77,57,141
Ethiopia	—	4,22,51,031

In terms of the percentage of national population of people without access to safe water the global situation is as follows:

Papua New Guinea	—	60%
Equatorial New Guinea	—	52%
Angola	—	51%
Chad	—	49%
Mozambique	—	49%

The report points out 'poor management of water resources as the biggest problem while mentioning how over-extraction of ground water aggravated the crisis. Aquifers provide 85 per cent of drinking water but levels are falling in 56 per cent of the country. Hand pumps are exacerbating the crisis in many areas by depleting the shallow aquifers. Misappropriation in planning and execution of water supply projects is another key factor'. The report further said that 'because of this water crisis, communities fall back on a single or distant source of drinking water, often leading to disputes'. Indeed, the situation of water scarcity has become so bad in some parts of India that acrimonious feuds in the population are not uncommon.

As a consequence of continuing droughts in parts of Maharashtra since 2014, the shortage in the availability of clean water is so acute that clean water is distributed through tankers on limited number of days in a week. The number of tankers deployed for the purpose are, however, grossly insufficient to deal with the demand of 5,00,000 people (of Latur town). Fearing possible occurrence of violence in the scramble of people grabbing water from limited number of tankers, the government was obliged to issue prohibitory orders in March 2016, that banned the assembly of five or more persons at six water treatment plants in order to prevent untoward incidents (an euphemism for fear of large scale uprising of people) as the residents struggle for water. A tense situation prevailed in the town with the police being deployed to ensure that there is no water theft.

Knowledgeable persons opine that the adverse effects of Nature's wrath have been compounded by human folly of growing water guzzling crops of sugarcane to feed the sugar mills. The crops in the region depend solely on ground water.

The report of the WaterAid also shows that in the developed world, a standard water bill is merely 0.1 per cent of the income of the people earning minimum wage; in India, however, anyone relying on a private water vendor (tank) would end up paying around 17 per cent of their daily income to get daily minimum supply of water.

Mismanagement of Ground Water Resources: Cultivation of Water Intensive Crops

The primary cause of ground water lowering is its over exploitation for irrigation in agriculture and horticulture. It is estimated by experts that in the last four decades roughly 84 per cent of the total addition to the net irrigated area has come through ground water. Latest estimates indicate that India is using 89 per cent of the ground water extracted for irrigation, 9 per cent for domestic use and 2 per cent for industrial use.

Closer review of the factors that have aggravated the water scarcity reveals that the cropping pattern is decided not on the sustainability of the crop within the available natural resources, including water for irrigation. Decisions with regard to the crop(s) to be cultivated are often influenced by extraneous considerations, without regard to the long-term sustainability of the most important ingredient for any cultivation, namely water. The cultivation of sugarcane in the water deficient districts of Maharashtra is the unfortunate victim of human misadventures. The first step towards establishing many sugar mills in such districts, in defiance of the concepts of techno-economic feasibility, created the Domino Effect which brought misery to the farmers, a few of whom committed suicides to escape the societal indignity for being unable to repay the huge loans they had taken to dig deep bore wells and install powerful electric/diesel pumps to draw underground water to irrigate their sugarcane plantation. Sugar mills needed sugarcane – farmers were lured to cultivate it – the crop required water – water was snatched out from the bowels of the Earth through deep and deeper bore wells – farmers took loans – even the little monsoon rains failed year after year – ground water level receded as far below to be out of reach for many bore pumps due to over exploitation. The society cannot remain a mute spectator any longer to this worsening scenario. Though the governments are quick to announce/offer financial relief to the affected families to earn political mileage, the problem needs to be tackled through scientific management of water resources, as a permanent eco-friendly solution lies elsewhere.

There are other examples where misplaced incentives have resulted in illogical choice of crop to be cultivated in surface water deficient areas. In order to protect the interests of the farmer, the government sets the 'minimum support prices' for the purchase of crop products by the traders from the farmer. A high-level committee appointed by the Government of India in 2014 observed that even though the minimum support prices are currently announced for 23 crops, they create a highly skewed incentive structure in favour of wheat and paddy (rice). Both of these crops are water intensive and depend heavily on irrigation (ground water in some parts of India which do not have an access to canal water). The result is that farmers began to prefer their cultivation over traditional, eco-friendly less water intensive crops, like pulses and oil seeds, even as the Indian variants of paddy, wheat, sugarcane and cotton plantations consume almost twice the amount of water to grow them as compared to China and the USA. Table 4 shows the average amount of water, in cu. m/tonne (1,000 kg) needed to grow different crops.

Table 4

Water use for crop production in different countries, cu. m/1,000 kg

Crop	*Brazil*	*India*	*China*	*USA*
Rice	3,082	2,800	1,321	1,275
Sugarcane	155	I55	119	103
Wheat	1,616	1,654	690	849
Cotton	2,777	8,264	1,419	2,535

Source: National Water Footprint Account, UNESCO-Institute for Water Education, May 2011.

In order to promote the cultivation of less water intensive crops, e.g. pulses and oil seeds, in lieu of wheat and paddy, the high-level committee suggested better price support for them as would motivate farmer to adopt their cultivation in preference to wheat or paddy. This would help to cut down the import of pulses. Whilst there was a quantum jump in the production of pulses around the year 2010, their production levels have since then remained stagnant. (http://www.ipga.co.in/pulses-trade-import-export). India has also been importing ever increasing large quantities of edible oils. In other words, there is sufficient market incentive to grow these crops in lieu of wheat. But India has been exporting wheat to a number of countries perhaps because farmers were lured to cultivate wheat. Does it not mean that we export large quantity of water, packaged in the form of wheat even though the water could have been used more fruitfully to grow crops that are otherwise being imported?

The society cannot any longer remain a mute spectator to this worsening scenario of water availability. Though the governments are quick to announce/offer financial relief to the affected families, the problem needs to be tackled through scientific management of water resources and promoting water literacy.

In the same vein, Pakistan has also been mining its underground water in their fertile plain of Punjab to cultivate the three water guzzling crops – cotton, sugarcane and wheat – the exports of which are the backbone of Pakistan's national economy. The crops are irrigated with both ground water (blue water) and surface water (green water). Their water table is continuously going deeper down; observation wells showed a fall between 1982 and 2000 from one to nearly two metres a year. There is acute shortage of drinking water in some parts. Yet exports continue as the country awaits ecological disasters on account of deepening water tables. The situation in their Baluchistan province is worse as underground water in some basins is said to have been exhausted, 'leaving their irrigated lands, outside the canal irrigated land, barren'.

Societal Intervention Against Use of Ground Water by Soft Drink Company

One specific case which drew worldwide attention relates to Coca Cola, which has its presence in a number of districts in India. One of their plants is located in Plachimada (Palakkad district) in Kerala. The Village Council gave a licence to the company to commence production in 2000, using ground water. The company started drawing around 5,10,000 litres daily of underground water and generated a large amount of polluted waste water which was openly discharged without any treatment.

The Company is located in an important agricultural region which is located in the rain shadow, drought-prone area of the Western Ghats. Yet, the area is regarded as the 'rice bowl of Kerala' due to cultivation of paddy, based on ground water irrigation. The cultivation was sustainable until the Company started its operations. As a consequence of huge withdrawals of ground water by the Company, the ground water table receded and, thereby, created severe water shortages for the agricultural community. In 2003, women from the village protested that their wells had dried up because of the over exploitation of ground water resources by the Coca Cola Plant. The villagers further alleged that the discharge of polluted water had also affected the quality of their underground waters.

The problem of pollution due to solid wastes came to international attention through the British Broadcasting Corporation (BBC) report. On 25 July 2003, the BBC reported the presence of heavy metals, lead and cadmium, in quantities higher than the approved limit in the sludge discharged by the Company with the waste water which the Company claimed was a fertilizer.

In April 2003, the Village Council revoked the licence stating that 'the excessive exploitation of ground water by the Coca Cola Company is causing acute drinking water scarcity'. This order was challenged by the company before the High Court of Kerala. The High Court observed that even without the existence of regulations for ground water exploitation, the existing legal position was that ground water is a public trust and the state has a duty to protect it against excessive exploitation. Additionally, it observed that ground water exploitation by landowners can result in negative environmental consequences. However, on appeal, the two Judge Bench of the High Court asserted the primacy of landowner's control over ground water in the absence of a specific law prohibiting extraction. The case is now pending in the Supreme Court.

Impact of Water Table Lowering on Vadose Zone

Another matter of great concern is the natural phenomenon that along with the lowering of the water table, the vadose zone in the (water) unsaturated soil is lowered as well. The fall of the vadose zone could take it beyond the reach of the root system of the trees on the surface; as a consequence, the trees begin to dry up, even as the other smaller vegetation had already dried up. Recalling the scientific conclusion that trees act as anchors to hold the clouds and to cause them to rain, the land, deprived of its normal supply of rain/surface water, turns towards desertification. The cattle grazing fields are lost. Instead of rains lashing the area, the land surface is frequented by dust storms. The runoff water from whatever rain that falls, easily erodes the surface of its silt and carries it to the river, and ultimately to one of the water reservoirs built in its downstream. The local ecology is eventually badly affected or destroyed.

A Peep into Global Scenario

Hydrogeology experts recognize that serious ground water depletion has already occurred in many parts of the world, notably in North Africa, the Middle East, South and Central Asia, north China, Mexico and arid/semi-arid regions of USA and Australia. In many areas, the water tables have fallen so low that wells can no longer produce water or the cost of pumping out has become prohibitive. It is almost truism to state that any country that has unabashedly used its ground water resources, without caring for their recharge, has had to face its effects on surface water flows and local ecology, e.g. the drying up of oases in Saudi Arabia as a consequence of heavy ground water withdrawals; or reduced flows in some rivers including in the United States and India. In some parts of India water tables have dropped more than 1 m. In areas of northern China, the water table has been dropping as fast as 1.5 m a year for the last 10 years.

Whilst the water tables in the industrial and developed European countries in the temperate climatic zone may have remained almost stabilized over the years, those of agriculture predominant developing countries have witnessed sharp drops. Many regions in North Africa or in the Middle East are experiencing harsh water shortages either on account of drying up of the aquifers or difficulties in pumping out from their decreased stocks. Though denied by the government, newspapers in Iran reported that 656 villages in the north-eastern province of Khorassan had to be evacuated about a decade back because there was no water supply to support the population. Iran was facing acute water crisis due to severe drought that had hit the country for the third consecutive year amid reports of wasteful water consumption. One estimate reported that Iran is over-pumping its aquifers by an average of 5 billion tonnes of water per year and the water table was falling by 2.8 m a year in the late 1990s.

Yemen is, perhaps, the only country in the world which has already come to the verge of bankruptcy with regard to water availability but the situation is not beyond redemption. According to a report prepared by the United Nations Development Programme, Yemen is using its water resources far faster than it can replenish them. The water tables have been falling, on an average, 2 m per year across the country. Gloomy forecasts of some futurologists suggest that its capital city, Sana'a, situated at an elevation of 2,250 m, could soon run out of its ground water supply. The spectre of a country running dry looms over Yemen's 25 million population, nearly 45 per cent of who live below the poverty line. According to World Bank estimates, per capita annual water consumption stands at a meagre of 120 cu. m, against a value of 1,000 cu. m regarded as the minimum by the World Health Organization. Experts on Yemen affairs believe that the dismal water situation is largely responsible for continuing internal violent conflicts and frequent droughts which have dealt a severe blow to its previous agricultural prosperity. A further blow to water availability for potable purposes was dealt when cultivation for the water intensive *'khat or qat'* was preferred over coffee. (Khat leaves are locally consumed as a chewable narcotic but coffee was largely exported). The irony of the situation is that Yemen is better endowed by Nature in terms of rainfall than other countries on the Arabian Peninsula. Whilst its rainfall at national is 135 mm/year, it is 520–760 mm in mountain areas to a high of 1,000–1,500 mm/year in the western highlands. The solution of its water woes lies in water harvesting and scientific management of its water resources.

According to information published by the Earth Planning Institute, 18 countries of the world, having a total population of over 3.6 billion people, are involved in over pumping of the aquifers. Amongst them are the most populous countries (China, India, Pakistan, Mexico) or all the desert, petro rich countries or the technologically advanced countries like the USA.

Risk of Water Wars Rises with Scarcity of Water

Water scarcity has the potential of triggering global conflicts. But the issue is linked to larger issues of population growth. Many years ago, Malthus, an 18th century British author had enunciated that: 'The power of population is so superior to the power of the earth to produce subsistence for man, that premature death must in some shape or other visit the human race'. Many global experts believe that the risk of conflict would grow as water demand is set to outstrip sustainable current supplies by 40 per cent by 2030. Internationally, 780 million people lack access to safe drinking water, according to the United Nations. By 2030, 47 per cent of the world's population will be living in areas of high water stress, according to the Organization for Economic Co-operation and Development's Environmental Outlook to 2030 report. 'Access to safe water is a fundamental human need, and therefore, a basic human right', said Kofi Annan, Former UN Secretary General.

The world's water supplies should be as much as to assure that every member of the population would have enough water to cover their personal and domestic needs. Water scarcity is an issue exacerbated by demographic pressures, climate change and pollution. The areas where water scarcity is the biggest problem are some of the same places where political conflicts are rife, leading to potentially explosive situations. The situation is no doubt triggered by heavy increase in population. The UN studies project that 30 nations will be water scarce in 2025, up from 20 in 1990. Eighteen of them are in the Middle East and North Africa, including Egypt, Israel, Somalia, Libya and Yemen.

Western Asia is in a state of turmoil ever since the USA toppled down the regime of Saddam Hussein in Iran. It is a strange coincidence that since then, the region has been experiencing acute water shortages due to wide failure of rains, particularly in the headwaters of the Tigris and Euphrates rivers in eastern Turkey. Consequently, the region has been witnessing one of the worst droughts in decades, which experts believe is fuelled further by the drying trend of the regions around the Mediterranean Sea on account of global warming. Many believe that the political unrests in Syria, Iraq and Yemen have been greatly influenced and facilitated by the scarcity of water – farmers abandon their desiccated fields and flood the cities with protests and thus become easy fodder for the political insurgency. Water bodies like rivers, canals, dams, and desalinated plants are all targets of attack by feuding so-called terrorists and governments of the region in the semi-arid region. Control of water supplies gives strategic control over both cities and countryside. Meanwhile, farmers are abandoning their field and moving en-masse to urban areas in search of employment, food and water.

Water is essential fodder in this conflict in which all major powers of the world are gradually being drawn in – the USA, Russia and the European Union – because of their vital interests in the rich petroleum deposits of the region. The mass entry of refugees from this region to the European Union in 2015–16 has given incentive for the involvement of global military powers.

Could there be a connection between global warming, water scarcities and conflicts in the region, and large population growth and rise of terrorism in the region which is now threatening to involve heavily armed military powers? A short answer is 'yes'. But the silver lining in the gathering dark clouds is that the situation can be averted by harvesting every drop of rain that falls on the parched lands. But for this water literacy must spread down to the common man.

References

Aquifers & Ground Water; US Geological Survey Survey; http://water.usgs.gov/edu/earthgwaquifer.html 10. (Retrieved, January, 2017.

Birzeit University: Groundwater Engineering: Chapter 1 - Occurrence of Groundwater http://www.hwe.org.ps/Education/Birzeit/Groundwater Engineering/Chapter%201-%20Occurrence%20of%20groundwater.pdf

Coastal Aquifers; Groundwater at Sea, Oilpro. http://oilpro.com/post/27309/coastal-aquifers-groundwater-sea.

Colarado Geo., http://www.douglas.co.us/water/water-supply/what-is-an-aquifer/.

Freshwater Submarine Springs of Bahrain: Threatened Heritage, http://www.academia.edu/2050947/2010_Fresh-water_Submarine_Springs_of_Bahrain_Threatened_Heritage_-_Poster_.

Ground Water and its Susceptibiity to Degradation, UNEP: http://www.unep.org/dewa/Portals/67/pdf/Groundwater_INC_cover.pdf

Groundwater Depletion, USGS. https://water.usgs.gov/edu/gwdepletion.html, logical Survey, USA

Kansas Geological Survey, Original Publication, 1965, Web Version created in 1998; URL: http://www.kgs.ku.edu/General/Geology/Sedgwick /gw01.html;

Kansas Springs: Kansas Geological Survey, Public Information Circular 11, http://www.kgs.ku.edu/Publications/pic11/pic11_1.htm.

Mr. Phillip's GCSE Geography, Battle for Biosphere, https://sites.google.com/a/kgv.hk/kgv-gcse-geography/home/battle-for-the-biosphere

Nelson, Stephen A., Tulane University, Metamorphic Rocks, Tulane University, www.tulane.edu/~sanelson/.../metamorphic.htm

Nelson, Stephen A., Tulane University, http://www.tulane.edu/~sanelson/eens1110/groundwater.htm)

Outline of Groundwater Hydrogeology: Meinzer : US Geological Survey Paper 494 https://pubs.usgs.gov/wsp/0494/report.pdf

Pioneers in the Pearl Trade: The Arabs, (Posted October, 2012) http://blogs.houseofgems.com/index.php/2012/10/pioneers-in-the-pearl-trade-the-arabs/

Porosity: https://en.m.wikipedia.org/wiki/Porosity.

The Global Water and Nitrogen Cycles, http://www.globalchange.umich.edu/globalchange1/current/lectures/kling/water_nitro/water_nitro.html

CHAPTER 6

Water in Indian Rivers

India is a country of many languages, but her diverse ethnic cultural traditions are unified by her rivers. The rivers are the heart and soul of Indian life. They are the pivot of India's culture and an aura of mystical divinity surrounds almost all major rivers. With the exception of the Thar Desert, the Indian land mass is covered with a number of large rivers, and the entire geographical area is dissected with their tributaries. Apart from influencing the social and cultural ethos of people, the rivers play an important role in the lives of the people by providing water, albeit now after cleaning, for potable purposes, irrigation and industrial uses and for providing an alternate method of transportation to railways and roads. Their waters are increasingly being used for the generation of eco friendly hydro electricity.

The alluvial basins of the Indian rivers as well as those in other countries, like of Indus (Sindhu) in Pakistan, or Yangtze, Salween Nhu and Mekong in other parts of Asia are the most populous areas in the world. Yet, it may sound a matter of profound revelation that the total quantity of water, contained in all the rivers of the world put together, is a mere 0.0001 per cent of all the water available on earth, inclusive of the oceans. The key for this minuscule quantity continuing to support modern man and all his activities lies in the facts that the habitat time of water in its river habitat is short (about 10 days) and this is ever recycling itself through the hydrological cycle. The hydrological cycle is Nature's way of doing large business having fast turn over with small investment. However, strong warning indications are now visible at the horizon that in very near future it may not be possible to sustain the increasing demands of fresh water.

Major River Systems of India

The river systems of India can be classified into four groups, viz.

- Himalayan rivers
- Rivers originating from the Central Indian Highlands
- Rivers originating from the Western Ghats. These are sub-divided into (a) eastward flowing rivers of the Deccan Plateau rivers, and (b) westward flowing coastal rivers, emptying into the Arabian Sea
- Rivers of the inland drainage basin, prominently in Rajasthan and Ladakh

All rivers are fed by rain water, either directly or indirectly. Most of their rain comes from the south-eastern monsoon winds prevailing over the Indian sub-continent for about four months after the harsh summer and at few places in south India it is supplemented by winter rains. However, the Himalayan rivers are additionally sustained by water from the melting of glaciers in summer months; this makes them perennial, according to natural plan of action. The rivers often receive enormous quantities of water within a short time in the monsoon season; consequently they swell their banks and may inflict ravaging flood fury on villages and towns in their path. Occasionally there are devastating flash floods. These rivers are getting increasingly sensitive for flash floods as a consequence of global warming. Kosi river (or the cursed one) in Bihar state is so-called because of its devastating floods and Damodar river earned the epithet 'Sorrow of Bengal' for itself for the same reasons. The rivers of the Deccan plateau are mainly fed by rain water but are not prone to floods on account of the nature of their terrain.

During summer, the flow in almost all rivers is greatly reduced, and some of their tributaries may even dry up, only to be revived in the next monsoon.

All river valleys have alluvial soil, are fertile and are densely populated. The Himalayan rivers carry huge load of sand and silt and detritus; consequently the flowing waters have enormous power to erode the surfaces in their path; they flow on soft bed. During floods, they spread the fertile alluvium far and wide. It is for this reason that the otherwise devastating and furious river, Kosi, is also worshipped as *Mother*. On the contrary, the rivers of the plateau flow on hard rock bed and do not carry much sand and silt. While coastal rivers, mostly of short lengths, gush down the peaks of the Western Ghats into the Arabian Sea in torrents during the rains. Their flows also slow down after the monsoon. Streams in the inland drainage system of Rajasthan are seasonal and partly saline in character, like the

Sambhar and drain into salt lakes. In the Rann of Kutch, the only river that flows through the salty desert is the Luni.

In modern times, waters of rivers from water surplus areas are reaching far off regions in water-deficient areas through long distance canals, such as the Indira Gandhi Canal to dry Thar region in Rajasthan or the Narmada Canal to Saurashtra in Gujarat. There are many other long distance canals. Wherever feasible, the water in the river is harnessed to produce eco-friendly hydroelectricity.

Monsoon Rains: The Primary Source of Renewable Water of Rivers

Irrespective of whether the rivers originate from the melting water of glaciers or by collecting the waters of large number of small streams running down the mountain slopes in the Himalayan or in the Western Ghats, or from springs or lakes, the primary source of their waters in all cases is rain. Recalling that rain clouds form from the moisture introduced in the atmosphere through evaporation of water from the surface (mostly) of the seas, the inherent scientific sense is also conveyed by stating that the Sun is the very primal source of all rivers. Hence, there is growing concern amongst climatologists for global warming which has already begun to seriously alter the pattern and amount of rainfall. Scientists attribute this change in the behavioural pattern of rainfall to global warming interfering with the natural processes of heat flow which had defined the traditional climate of any place. In this context, it would not be a misnomer to call all the rivers as 'Suryaputrees' or daughters of the Sun. In Indian mythology this epithet is, however, reserved for only two prominent rivers– for Yamuna, originating from the Yamunotri glacier up in the Himalayan ranges, and for Tapi or Tapti, originating from a pond near Betul (MP) at the eastern end of Satpura mountain ranges in central India.

Nature had designed Yamuna to be a perennial river and Tapi, an essentially monsoon influenced river. Both carry enormous quantities of water after the monsoon rains but Tapi nearly dries up some time after the rains. Vast stretches in the flow path of Yamuna now often run dry, particularly in summer months due to human interference with the flow of its waters.

The stark fact of monsoon rains being responsible for Tapi is engrained in Indian ethos as the inhabitants of Tapi catchment area celebrate her birthday every year on the 7th day of the month of

Assaadth (Indian Almanac), heralding the onset of monsoon, by organizing fairs at several places along its banks (The word 'tapi' or 'tapti' itself means *heated).* They hold Tapi in great reverence and regard her as a great redeemer by proudly claiming that if Ganga washes the sins away of those who take a dip in her holy waters or Narmada does so of those who have her 'darshan', Tapi does so if one simply invokes her in his/her mind. Such is the hold of the river Tapi or Tapti on Indian ethos.

Figure 1

Schematically showing the geologically older mountains and plateaus influencing the direction of flow of rivers

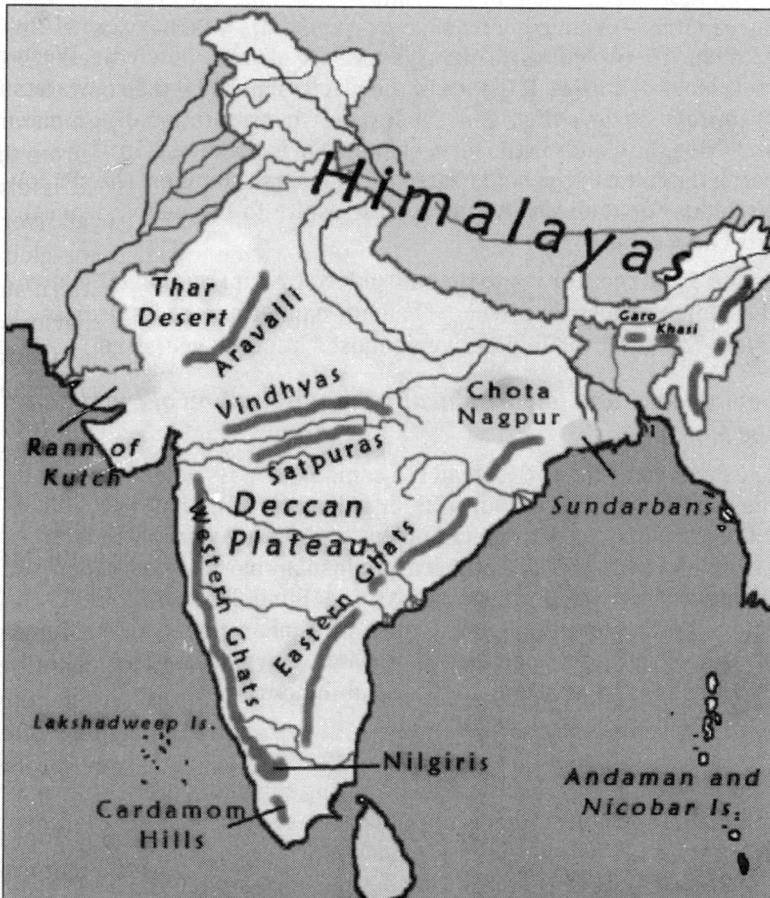

How does then one reconcile the fact of two Suryaputrees having two vastly different sources of their respective origin? We are now attempting to solve this enigma scientifically. One fascinating and enduring feature of Indian thought and culture is that complex facts of Nature/science are communicated to the masses through simple stories which can be easily comprehended by the common masses. In mythology, the Creator entrusted the task of engineering the nascent planet earth into a habitable one to Lord Vishwakarma who had his twin daughters (one named Sanjana - meaning the gentle creator, suggestive of sunlight and the other Chaayaa - or the comforting shadow) married to Sun god Suryadeva. After all, both 'light' and 'shadow' are manifestations of the presence of Sun. Where there is light, there must also be a shadow. Yamuna and Tapi are both Suryaputris - Yamuna of Sanjana - and Tapi of Chaayaa, in this context. The two consorts of Suryadeva are always shown together in mythological stories. If the ice of the glaciers melts by the gentle heat of sunrays, so does their heat cause massive evaporation of sea water and drive the clouds in the form of monsoon in the direction of India to parch the thirst of the land heated up by a harsh summer. The shadow of clouds (or their *chaaya*) provides respite to the people from the scorching heat.

A great recognition to the role played by Tapi river in influencing the Indian ethos came when Thailand named one of its prominent rivers *Tapi* after the Indian river, almost a hundred years ago.

Geologically and Geographically Unique Position of India and the Monsoons

The Peninsular Deccan Plateau is a composite mass of land formed by the coming together of four different blocks of land masses that have different plate tectonic, metamorphic and paleogeographic histories. These blocks are: The southern peninsular block, often called the Gondwana or the Dharwar Block; The Bundelkhand Block, Trans Aravalli Block and the South Indian-Sri Lankan Block. These blocks have been brought together by tectonic movements. The Dharwar Block itself is an assembly of three, sub-blocks. These sub-blocks are: Karnataka, Bastar and Singhbhum.

The rock formations of the Deccan are greatly influenced by the heat of the magmas and the accompanying seismic events (geologists call them tectono-thermal events). These have changed the character of the igneous rocks into those of metamorphic rocks and introduced several major faults in the strata, besides forming infinite number of minor faults and fissures and gaps in the rocky strata.

The process of converging of the sub-blocks together resulted in their 'suturing' with each other along which the present Pranhita-Godavari and Mahanadi valleys are formed (in geology, a suture is a joining together along a major fault zone, of separate terrains or tectonic blocks. The suture is often represented on the surface by an orogen or mountain range).

The unique topographical relief of India and its geographical position on the globe is, therefore, entirely due to geologic processes that were set in motion several hundred million years ago due to the fragmentation of the massive supercontinent of Gondwanaland into a number of plates and the slow northward drift of the Indian-Australian Plate/s towards the much larger Eurasian Plate. When the two plates finally collided about 50 million years back, a part of the Indian Plate got subducted under the Eurasian Plate. The subduction of the plate caused disappearance of the Tethys Sea, formation of the folded Himalayan Mountain ranges and of the Tibet Plateau. It may be recalled that subduction is a process that takes place when one tectonic plate moves under another tectonic plate. The process of tectonic shift is continuing to occur in contemporary times, albeit at very slow rates, about 2 cm per year as a result of which both Himalayas and the Tibet Plateau are rising. The subduction zones are sites of high rates of *volcanism*, *earthquakes*, and *mountain building*.

Three geological morphologic features of India have defined the flow of rivers. These are shown in Figure 1 and are:

- The Himalayas and the associated mountainous regions
- The massive Sindhu (Indus) – Ganga (Ganges) – Brahmaputra alluvial plain
- The Deccan Peninsular Plateau, which really is a confederation of several different plateaus brought together with each being independently formed in different geological events.

The geographical positioning and geological structuring of India are unique amongst the countries of the world as together they have made it a land of water and greenery and have prevented the spread of the Sahara Desert into it. It is surrounded by vast expanses of water bodies (Arabian Sea on the western side of the peninsular India, the Indian Ocean on the southern and the Bay of Bengal on the eastern).

The Deccan Peninsular Plateau bounded by the Western Ghats and the Eastern Ghats. In Hindi *ghat* means 'river landing stairs' or 'mountain pass' and has been accepted in English language in its plural form to mean the mountains themselves. The Western Ghats are flat-topped hills presenting the stepped or terraced topography; hence the name. The 'stepped' topography is largely due to the Deccan Plateau being formed through successive and massive ejections or extrusions of igneous lavas from the womb of the earth which on reaching the surface of the Earth melt (on account of accompanying decompression caused by the loss of pressure head) and spread horizontally to form the successively bedded strata, which on weathering, have given them the surface relief in the characteristic form of 'landing stairs', hence the name Ghats.

The Deccan Plateau is the geologically the oldest and seismically the most stable part of the Indian landmass lying between the Arabian Sea and the Bay of Bengal and includes a substantial portion of the area to the north of the Vindhyan Range. The plateau is 300–750 m above the sea level and generally slopes towards the east. Its hills ranges have been subjected to heavy erosion. The main peninsular block is composed of granites, gneiss, granite-gneiss, and schists, as well as of more geologically recent basaltic lava flows. They may be the primary igneous rocks or their metamorphosed versions. Their names are mentioned to excite the minds of those wishing to seek further knowledge. The strata/rocks composed of these minerals are not hospitable to water, being crystalline or fine grained solids and behaving as a 'true solid' in so far as the percolation of water through them is concerned; in substance, water cannot permeate through them. However, these rocks have been subjected to intense volcanic and seismic activity which has caused intense faulting; fracturing/fissuring, formation of escarpments, rifts is strata, etc. Geologists tell us that LANDSAT imagery confirms the presence of a large density of faults along the trend of the Ghats. In other words, the rocky strata have developed secondary porosity which forms an important habitat for rain water in the rocky strata of the Deccan Plateau.

Percolation of rain water through them to reach greater depths is a slow process. It is, therefore important to find a transit store on the surface of the Earth which can hold or capture the rain water and prevent its becoming a 'runoff'. Surface soils provide this 'transit store'. It is, therefore, important to prevent the loss of surface soil through erosion or mining activities and other forms of human interference.

The Western Ghats rise steeply as an escarpment of variable heights from the coastal plains of Arabian Sea, the highest elevation (1,430 m) being at Doda Betta (an escarpment is a steep slope an area of the Earth where elevation changes suddenly. It is formed through geological processes which cause faulting and result in the formation of two relatively flat areas but of different elevations). They contain several lateritic plateaus and peaks separated by passes and saddles. The southern part of the Western Ghats is separated from the main northern Sahyadri range by the 21–31 km wide Palghat Gap at an elevation ranging from 75 to 300 m while the bordering hills rise up to 2,000 m on either side. This is a sudden break in the continuity of this mountain range. It is through this gap that moisture-bearing clouds of the south-west monsoon penetrate some distance inland, bringing rain to parched Mysore region. Their eastern sides, however, slope gently.

The present day Western and Eastern Ghats are the eroded remnants of ancient chains of mountain ranges. When original mountains are eroded by natural agents of gradation, they become relict mountains. Gradation is the set of processes that work on surface rocks to break, loosen, move, and finally deposit the smaller pieces. It may be recalled that these processes are: weathering (the chemical decay or mechanical break-up of rocks), erosion, transportation (the removal of weathered rocks by moving water, wind, ice, or gravity), and deposition. The presence of water and atmosphere drive the chemical decay of rock. Gravity moves landslides. Moving water, wind, and ice create landforms such as river channels, deltas, beaches, sand dunes, glacial deposits and valleys. The Eastern Ghats now constitute a rather discontinuous series of low mountain ranges running almost along the coast of Bay of Bengal. The eastern flowing rivers enter the Bay of Bengal through gaps in the ranges. The Godavari River, for example, traverses through a 65 km wide gorge to form a delta with the Bay of Bengal.

The second structural unit of the Indian landmass between the Himalayas and the Deccan Plateau is the Sindhu-Ganga-Brahmaputra Plain. The Plain was built by the deposition of alluvium and other sediments in what was the Tethys Sea by rivers from the emerging Himalayas and those older ones flowing northward from the Deccan Plateau. The thickness of the alluvium deposits may extend up to about 2,000 m. The overall slope of the Ganga Plain is imperceptibly small at about 95 mm/km. In the eastern portion of the plain is the delta jointly formed by the Ganga and Brahmaputra rivers. Their deltaic area is characterized by annual flooding attributed to intense monsoon rainfall, an exceedingly gentle gradient, and an enormous discharge that the alluvium-choked rivers cannot contain within their channels.

The highest folded mountain ranges in the world at its northern boundaries – the Himalayas – obstruct the northward movement of the moisture laden monsoon winds. Along with the Western and Eastern Ghats, the Satpura, Vindhyas and Aravallis are also called the relict mountains. The undulating topography of the Deccan Plateau (which includes Chotanagpur as well) helps the monsoon winds to spread and cover the entire country, albeit selectively. Most of the country is, thus, swept by the moisture laden monsoon winds rising from the vast expanse of the water bodies. These features compel the winds to shed all their moisture on the Indian landmass and other South-East Asian countries as well.

Along with the Himalayas (and Karakoram ranges), the Deccan Plateau plays a prominent role in the making of water sheds from where the rivers originate. The munificence of the Himalayan water treasury supports the existence of almost half the population of the world distributed in several countries of the world, notably India, Bangladesh, China, Myanmar, Pakistan and Thailand.

What is a Plateau?

A plateau may be defined as that upland which has at least one side of very steep slope standing well above the neighbouring surface and whose upper part is extensive and almost flat. Generally, the mountains are more than 1,000 m high. The mountains, whose height is less than 1,000 m, are called hills. There could be several types of plateau. The Deccan Plateau of India, as described earlier, is a huge upland formed by massive layers of basaltic lava with the Western Ghats rising steeply at its western boundary. Geologists call it a volcanic plateau.

Plateaus can also be formed by the uplifting of landmass by violent geological activities in such a manner that the middle portion is raised and the sides are rounded; a good example of the dome-shaped plateau is the Chhotanagpur Plateau in India. The Tibetan Plateau is also formed by the uplifting of land due to tectonic movements is surrounded on all sides by hills and mountains; this type of plateau is designated as an 'intermontane' plateau.

These plateaus have defined, as we shall see a little later in the chapter, the pattern of water sheds and have given flow to some of most important rivers of India and the world. The physiognomy of the Deccan Plateau has, indeed, been determined by the intense seismic activity triggered by the push and collision of the Indian Plate with that of the Eurasian. As a consequence thereof, the Deccan Plateau is heavily vivisected with geological faults, rift valleys and suture zones between sub-plateaus - all of which are geologically weak zones where erosion of land surface by flowing water eventually succeeds in paving the way for rivers to flow.

Orographic Precipitation of Rain on the Himalayan Heights

After shedding their moisture over the land mass, the winds rise up to cross the Himalayas. The progressively increasing cooler conditions up there on the Himalayan ranges causes the residual moisture to condense in the form of tiny droplets of water which eventually fall as rain on the India facing sides of the mountain slopes. The meteorologists call it orographic precipitation, which is defined as the rain produced when moist air is lifted as it moves over a mountain range. As the air rises it gets cooled. Its capacity to hold water vapour decreases; excess vapour is condensed in the form of tiny droplets which form the orographic clouds which precipitate their water mostly on the upwind side of the mountain range.

As they rise up further to cross the Himalayas to continue their onward journey to the Northern Central Asian lands, the monsoon winds are obliged to shed even the last traces of moisture in the form of rain/snow because of very low temperatures prevailing at the Himalayan heights.

Impact of Topographical Relief on Distribution of Monsoon Rains

The Western Ghats (Sahyadri Mountain ranges) along the western coast of peninsular India, and the lesser Eastern Ghats together with the lofty Himalayas make the topographical relief of India unique. They make a sort of trough in which the monsoon clouds are held a little longer. In other words, the monsoon clouds stay over India for a longer period. With increased residence time, thus, available, monsoon winds are able to shed off a good amount of their water over the Indian land mass. Once trapped or anchored over the Indian landmass, the micro pattern of rainfall is greatly influenced by the local topographical morphology and there could be large variations of rainfall in any one region. It has been scientifically established that forest cover helps to 'anchor' the clouds.

On account of the undulating topographical features of India, the spread or distribution of monsoon winds is not uniform across the vast Indian landmass. Therefore the rainfall is not equitably distributed across the country. If there are regions in Meghalaya with very high rainfall (more than 11,000 mm per annum at Mawsyram and Cheerapunji), there are also dry or very dry regions in Saurashtra and Rajasthan where the rainfall could be less than 400 mm per annum.

On the wind facing side or the upward wind side of the Western Ghats, i.e. towards the Arabian Sea, there is very high rainfall due to orographic precipitation. The rain shadow area falls on the eastern side of the Western Ghats and overlooks the peninsular India. Thus some districts of Maharashtra are perpetually rainfall deficient. They often face drought-like situations. These regions are more dependent on ground water than on the surface water resources for both domestic and farming needs.

In recent years, shallow bore wells were replaced with deep and deeper bore wells which led to excessive withdrawal of ground water through powerful pumps. As has been explained in a previous chapter, heavy withdrawals of water from a single unit powerful pump play havoc with surface ecology and local surface water resources. Therefore, there is a strong need to locate the bore wells in the context of local geo-hydrology.

Annual Rainfall in the Country

Experts have estimated that if the annual rainfall over the country for the past 110 years is considered, the average rainfall for the entire country, taken as a unit, is 899 mm per year but its statistical fluctuation could be as large as 20 per cent. Experience shows that if the rainfall exceeds by about 10 per cent, there could be severe floods in certain regions and a 10 per cent shortfall is suggestive of drought. Table 1 provides data about the rainfall in the country arranged according to the prominent rivers along with some information about surface water man made resources for fresh water. Table 1 also gives the status of total available rain water as well as that of 'runoff' water.

Table 1

Main basins or drainage areas: their areas, water projects and surface water available for recharge average monsoon rainfall and average monsoon runoff, committed storage of surface

	Basin	Drainage Area sq. km	Average Monsoon Rainfall cm/year	Average Monsoon Runoff (a) Million cu. m Million cu. m	Definite Storage Capacity (b) Million cu. m	Available Run off c = (a–b)
1.	Sindhu	3,21,289	32	58,640	16,992	29,837
2.	Ganga, Brahmaputra Barak	10,97,540	111	8,78,800	1,50,963	5,87,120
3.	Godavari	3,12,812	94	1,07,080	36,248	47,437
4.	Krishna	2,58,948	60	61,010	37,230	4,649
5.	Kaveri	87,900	75	18,860	8,017	6,151
6.	Pennar	55,123	48	6,170	2,519	2,152
7.	Maha- Goda + Krish-Pen*	74,350	66	15,250	3,824	8,563
8.	**	1,00,140	90	15,950	1,441	12,410
9.	Mahanadi	1,41,590	148	60,190	25,274	20,051
10.	Brahman Baitarni	51,822	140	32,600	13,129	11,616
11.	Subarnarekha	29,200	120	9,680	3,806	3,574
12.	Sabarmati	21,674	76	3,430	1,445	1,136
13.	Mahi	34,842	88	10,650	4,534	3,464
14.	***	3,21,874	48	13,590	5,838	4,350
15.	Narmada	98,796	100	36,860	23,202	1,851
16.	Tapi	65,145	70	16,200	11,093	000
17.	Tapi-Tandri zone	54,120	216	97,760	10,277	74,110
18.	****	58,000	187	80,320	14,271	53,022
19.	Internal Drainage Area in Rajasthan	26,900	22	-	-	-
20.	*****	-	114	24,800		-
	Total	65,78,72	1,905	5,61,960	2,19,141	2,84,373

* East flowing rivers between Mahanadi–Godavari and Krishna–Pennar

** East flowing rivers between Pennar and Kanya Kumari

*** West flowing rivers of Kutch and Kathiawar, inclusive of Luni

***** Indian portion of rivers draining into Bangladesh and Myanmar

Source: Central Water Commission.

The monsoon also covers other countries of South-Asia and South-East Asia. Its Indian quota varies from year to year. Because

India is traditionally a predominantly agricultural country, where farmers depended entirely on rains or rain fed ponds and shallow dug wells, information in advance about the onset of monsoon and its strength helped the farmer to prepare his fields for the sowing season ahead. Different methods were and are used to predict the time for the onset of rains (discussed later in the chapter).

Rivers and Their Basins

Rain water falls all over the surface of the earth. A part of it immediately percolates down to replenish the ground water resources, a part is evaporated and the rest finds its way to large rivers. The natural topography of the land helps to collect rain water into small streams or rivulets which deliver it to larger rivers.

If one were to look at India from high up in space, its undulating topography would look like a comprehensive tapestry built with mountains, valleys, ridges and plains, themselves being made of a number of basins – one fitting into another. In other words, a large basin accommodates a number of smaller sub-basins; of all possible shapes – some circular, some oblong, some green, some barren but always a number of them comprehensively fitted into a larger basin in the style of a three dimensional jig-saw puzzle. In the deepest portion of each basin, would be spotted a pond or a rivulet which collects the rain water flowing down the slopes of the basin with the rivulet flowing in the general direction of the slope of the basin; therefore the local slope of the surface gives direction to the flow of the rivulet or river. The smaller rivulet submits itself to larger rivulet from another sub-basin; the basin of the smaller rivulet now becomes another part of the basin of the larger rivulet with which it has mingled its own identity. In other words, the larger rivulet/river grabs the basin/s of the smaller ones. The process of smaller rivulet merging with the larger one continues till a big river is formed; all the sub-basins integrated together become the comprehensive basin of the large river. In this context, a basin is the catchment area of the rivulet/river. The rivulets in very many micro basins in the alluvial plains are seasonal, being active only in the rainy season but those in the plateaus may last a little longer as underground water may continue to seep out and thus constitute the base flow of the river. Whilst the basin is the catchment area for its rivulet, the rivulet itself can be regarded as draining the basin of its water; in this context, the basin is often referred to as the drainage area or regarded as a basic

Figure 2

Rivers and Basins (in the context of Table 2)

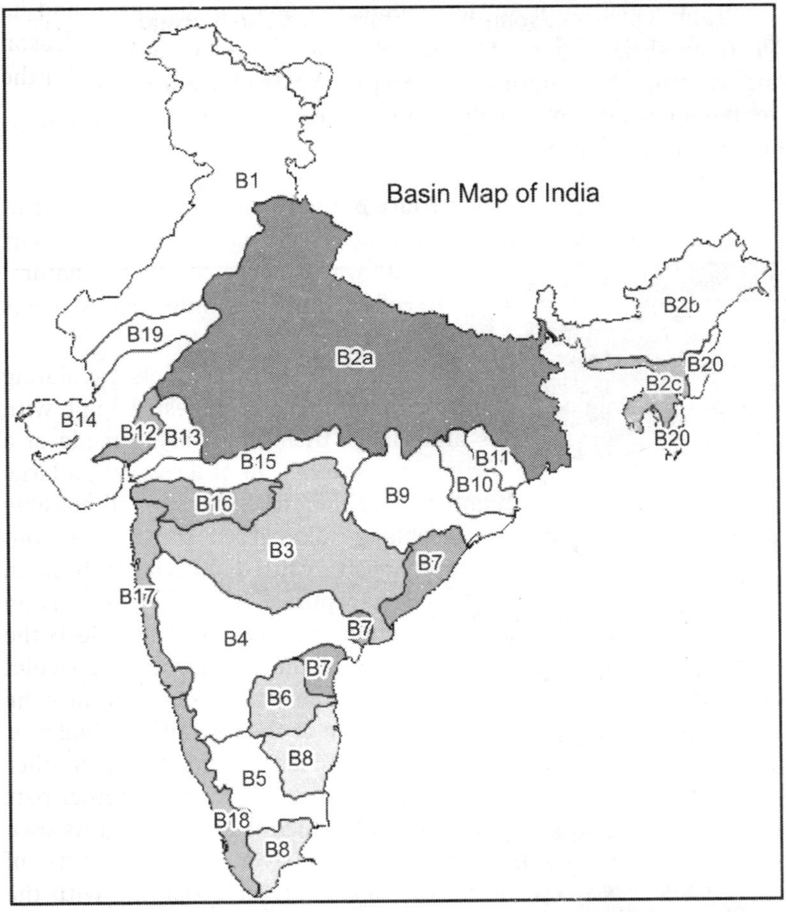

Source: Central Water Commission and Indian Space Research Organization.

hydrological sub-unit. The water may be said to be the 'runoff' water of the basin.

River basin is considered as the basic hydrological unit for planning and development of water resources. The entire country has been divided into 22 basins as per National Commission for Integrated Water Resources Development Plan (India). These basins are shown in Figure 2; their names and area are given in the Table 2. All major river basins and many medium river basins are inter-state in nature and cover about 81 per cent of the geographical area of the country.

Table 3 provides some basic information about prominent rivers. The Ganga-Brahmaputra-Meghana basin is the largest in the country, both in terms of its physical size and population which is nearly half of the Indian population residing in river basins. In terms of size, it is followed by the basins of Krishna, Godavari and Sindhu.

Table 2

Names of river basins and their areas

S. No	*Basin Code*	*Basin Name*	*Area (sq. km)*
1.	B1	Indus (Up to border)	3,21,289
2.	B2a	Ganga	8,61,452
3.	B2b	Brahmaputra	1,94,413
4.	B2c	Barak and others	41,723
5.	B3	Godavari	3,12,812
6.	B4	Krishna	2,58,948
7.	B5	Cauvery	81,155
8.	B6	Pennar	55,213
9.	B7	East flowing rivers between Mahanadi and Pennar	86,643
10.	B8	East flowing rivers between Pennar and Kanyakumari	1,00,139
11.	B9	Mahanadi	1,41,589
12.	B10	Brahmani and Baitarni	51,822
13.	B11	Subernarekha	29,196
14.	B12	Sabarmati	21,674
15.	B13	Mahi	34,842
16.	B14	West flowing rivers of Kutch and Saurashtra including Luni	3,21,851
17.	B15	Narmada	98,796
18.	B16	Tapi	65,145
19.	B17	West flowing rivers from Tapi to Tadri	55,940
20.	B18	West flowing rivers from Tadri to Kanyakumari	56,177
21.	B19	Area of inland drainage in Rajasthan	-
22.	B20	Minor rivers draining into Myanmar and Bangladesh	36,202

Whilst these statistical facts appear rather dry, their poignancy may be felt when they are recast to show the annual per capita water availability in the different river basins or hydrological units. This is done in Table 4. It is seen that the water availability in the southern regions of the country may be better than in the above mentioned two basins, yet is considerably less than the 1,000 cum/person/annum mark, which global experts regard as the minimum necessary to sustain life; a value less than this amount is regarded as creating 'conditions of scarcity'. In this regard, the narrow basins on the western side of the Sahyadri Mountains and in that of Narmada are in a happy situation. But again most of the surplus water of the former drains into the Arabian Sea.

Table 3

Details about some prominent rivers

	Name	*Length km*	*Source*	*Mouth*	*Spread of Basin*
1.	Sindhu	3,100	Kailash Mountain Tibet (Mansarovar)	Arabian Sea	India and Pakistan
2.	Ganga Bhagirathi	2,480	Gangotri Glacier	Sundarban, Bay of Bengal	Uttarakhand, UP Bihar Bengal
3.	Yamuna	1,370	Yamunotri Glacier	Prayag UP in Ganga	Haryana and UP
4.	Brahmaputra	725	Mansarovar Lake Tibet	With Ganga	North-east India
5.	Kaveri	805	Coorg, Western Ghat	Bay of Bengal	Karnataka Tamil Nadu
6.	Godavari	1,465	Near Nasik Western Ghat	Bay of Bengal	Maha, Telangana AP
7.	Krishna	1,400	Mahabaleshwar	Bay of Bengal	Maha. Telangana AP
8.	Narmada	1,312	Amarkantak MP	Arabian Sea	MP, Maha, Gujarat
9.	Tapi	724	Betul MP	Arabian Sea	MP, Gujarat
10.	Mahanadi	858	Amarkantak Plateau	Bay of Bengal	Jharkhand Odisha Chhattisgarh
11.	Vaigai	240	Cardmon Hills	Bay of Bengal	Tamil Nadu
12.	Periyar	244	Cardmon Hills	Bay of Bengal	Tamil Nadu Kerala
13.	Thamiraparani	123	Augustiyaar Hills	Gulf of Mannar	Tamil Nadu
14.	Sone	784	Maikal, Amarkantak	Tributary of Ganga	MP UP Bihar

Cont'd...

...Cont'd

15. Chambal	960	Mhu near Indore	Tributary of Yamuna	MP, UP, Rajs
16. Ghaghra	1,080	Mapcha-Chugu Glacier Tibet	Tributary of Ganga	UP and Bihar in UP
17. Sarayu	350	Confluence of Sharada and Ghagharaa	Tributary of Ganga	Uttarakhand, UP and (Nepal)
18. Sharada	350	Kalapani in Uttarakhand	Tributary of Ghaghra	Uttarakhand, UP and (Nepal)
19. Kali Sindh	470	Bagali, MP	Chambal	MP
20. Chenab	960	Bada Lachcha-La, Lahul Spiti, HP	Tributary of Sindh (Pak)	HP, J & K
21. Satluj ***	1,450	Raksha=sthal Lake, Kailash Mt. Tibet	Tributary of Sindh (pak)	HP in India

*** Highest rail bridge in the world; height 359 m; length 1,315 m in India.

Table 4

Per capita annual water availability in river basins

	River	*Population in Millions*	*Per capita Annual Water Availability, cu. m*
1.	Sindhu	58	1,271
2a.	Ganga	494	1,062
2b.	Brahmaputra	40	13,407
2c.	Meghana	9	5,667
3.	Godavari	74	1,486
4.	Krishna	84	933
5.	Kaveri	40	530
6.	Subernarekha	13	958
7,	Brahmani + Baitarni	13	2,113
8.	Mahanadi	37	1,826
9.	Pennar	13	472
10.	Mahi	14	761
11.	Sabarmati	14	263
12.	Narmada	20	2,253
13.	Tapi	20	731

Cont'd...

...Cont'd

14.	West flowing rivers between Tapi and Tadri	36	2,460
15.	Rivers between Tadri and Kanyakumari	45	2,528
16.	East flowing rivers bet. Mahanad and Pennar	33	692
17.	East flowing rivers bet. Pennar-Kanya Kumari	62	262
18.	West flowing bet Kutch Saurashtra and Luni	30	496
19.	Inland draining rivers of Rajasthan	10	
20.	Minor rivers draining into Myanmar and Bangladesh	2	14,976

Source: Central Water Commission data for 2010 (estimates).

We intend to present discussion of river basins as illustrative of salient points concerning management of water resources and not as an exhaustive presentation of their river systems for which reference may be made to a standard book on geography.

What is a Water Shed?

A watershed is an area of land that collects rain water and drains it off to a common outlet; the word is often used interchangeably with drainage basin or catchment area. The words 'common outlet' is a broad term as it could refer to 'sea' as the common outlet of water from a set of rivers; or to a river as the common outlet for its tributaries and sub-tributaries. Usually, geomorphologic features of the surface of the Earth are involved in deciding the area of the watershed. If we regard the Sindhu-Ganga Plain as the common outlet, then the 'water shed' covers the entire Himalayan range. In this manner, if we regard the Indian land mass as the common outlet of rain water, then there are three principal water sheds in the country. These are:

- The Himalayan Water Shed
- The Western Ghats Water Shed
- The Central Indian Highlands

The Himalayan Water Shed

The first and the most prominent water shed is the Himalayan region, largely on account of heavy rainfall and snow fall in the vast expanses of its valleys on the India facing slopes. Because of (i) orographic precipitation of rain and snow on higher elevations of the Himalayas, (ii) melting of snow, and (iii) melting of glaciers, in summer, most of the large Himalayan rivers are perennial. In fact, the source of the Yamuna river is the Yamunotri glacier and that of Ganga, the Gangotri glacier but there are other rivers which originate from the melting of glacier(s), such as the Sapt Kosi. Rivers from the northern slopes of the Himalayas on the Tibetan side also flow into India, such as the Indus (Sindhu) river or the Brahmaputra river.

The Western Ghats Water Shed

The Sahyadri or the Western Ghats constitute the second important water shed in the country. This water shed could be divided into two sub-sheds – one on the western sea-facing side and the other, the east-facing side overlooking much of the Deccan Plateau. Their character is vastly different from each other.

The western face of the Ghats faces the Arabian Sea and is, naturally, its monsoon wind-facing side. Although the heights of the Western Ghats are dwarfish in comparison to those of the Himalayan, their close proximity with the sea makes them an ideal venue to receive rains in abundant measure. Further, as the moisture laden winds rise above their heights in order to continue their onward journey, they cause heavy rainfall in their region. The normal rainfall is augmented by orthographic precipitation at higher elevations. However, much of the rain water goes back to the sea as the coastal plains between them and the sea are very narrow. There are, therefore, a very large number of short rivers, besides the generally known Zuari, Mandoi, Sharavati and Periyar.

The eastern side of the Western Ghats naturally is in their rain shadow area. It, therefore, sees much less rain than the western side. Yet, it gives rise to all the rivers of the Deccan Plateau – notable amongst them being Godavari, Krishna, Pennar and Kaveri, besides many others. In the Pennar and Kaveri basins, the per capita/annum water availability is less than the globally accepted value of 1,000 cu. m for a life without feeling the pangs of water stress. When this value falls below it, a situation of water scarcity is created which has already begun to result in inter-district and inter-state feuds on water related issues. Every drop of rain water – irrespective of whether it seeps into the ground or flows as surface water – is important as both are also the natural store houses of water.

Central India Water Shed

The central Indian highland, bounded on the western side by the eastern ends of Satpura and Vindhyachal mountains, Aravalli highlands and the Gangetic plain on the north but containing the hills of Ramgarh and Chota Nagpur and northern parts of the Deccan Plateau is the third important water shed of India. It is a unique kind of water shed as it houses the origins of several important rivers which flow from its tablelands in all directions; Narmada and Tapi to the west; Chambal, Betwa and Sone towards north, Damodar and its tributaries flowing almost to north-east, besides its largest, east flowing inter-state (Jharkhand, Odisha and Bengal) river, Subarnrekha; the South-east flowing Mahanadi and its tributaries and south flowing tributaries of Godavari river which itself has its origin in the Western Ghats water shed.

The character and behaviour of rivers is different from one another as they are influenced by the geology of the regions of their flow. Most of them with the possible exception of Narmada are basically ephemeral rivers; Narmada and Tapi flow gently, Damodar was known for the violence of its flow, until its flow and that of its tributaries was moderated through the construction of a series of dams and water reservoirs in the 1950s; Mahanadi and Sone, both owing their flow to the monsoon rain, are known for their widespread floods in the rainy season.

Main Drainage Zones or Basins

The Ganga Basin

Ganga is the most sacred of the Indian rivers. However, during the early Vedic Age of the Rig Veda, the Indus and the Saraswati river were the major sacred rivers, not the Ganga. But the later three *Vedas* give much more importance to the Ganga. The spread of the Ganga basin is vast – from the Himalaya and the Trans-Himalaya in the north, to the northern slopes of the Vindhya range in the south, from the eastern slopes of the Aravalli in the west to the Chota Nagpur Plateau and the Sundarban delta in the east. The Himalayan portion of the basin contains 9 of the 14 highest peaks in the world over 8,000 m in height, which besides the highest Mount Everest are: Kanchenjunga, Lhotse, Makalu, Cho Oyu, Dhaulagiri, Manaslu, Annapurna and Shishapangma.

The Ganga drains a total catchment area of 10.631 lakh sq. km, out of which 8.61 lakh sq. km lies in India. Thus, Ganga basin covers 26 per cent of geographical area of the country. The annual rainfall in the Ganga basin varies from 500 mm. to 2,500 mm in the mountains and

400 mm to 1,250 mm in the plains. Most of rainfall is concentrated during the months of June to September.

Whilst a significant portion of water in Ganga comes from the Himalayan mountain system, a good amount also comes from the rivers originating from the central Indian water shed.

The hydrologic cycle in the Ganga basin is largely governed by the south-west monsoon as about 84 per cent of the total rainfall occurs in the June to September period. Consequently, flow in the Ganga is highly seasonal and can cause floods.

It may be restated that several smaller (may otherwise be prominent) rivers, with their own basins, become tributaries of the main river, after the name of which the drainage zone is known, thus an extensive drainage zone is formed. In this manner, the massive and widely spread Ganga basin is formed through its amalgamation with the basins of its tributaries, most notably that of Yamuna. Thus, Ganga may only have been a small river in its exclusive basin in the higher Himalayan ranges but becomes a mighty river with the confluence of several other big rivers with it.

A comprehensive map of Ganga basin is shown in Figure 3. The figure shows some of the main rivers which become its distributaries, for example, rivers like Sharda, Gandak, Sharada and Sapt Kosi or simply the Kosi coming from the north and Yamuna, Gomati, Ram Ganga, Sone (and others) confluencing with her on the left. These tributary rivers have their own tributaries, like Yamuna having hers – Rishi, Hanuman Ganga, Tons, Rupin, Bata besides Chambal, Betwa, Ken to name just a few. Rivers originating from the hills of Chota Nagpur like Damodar and Sone from the Amarkantak hills also become tributaries of Ganga. According to geologists, some of the non-Himalayan tributaries of Ganga may be older than the Ganga itself.

As can be inferred from Figure 3, the Ganga basin covers almost all the north Indian territory and also extends into the neighbouring country – Nepal. When the basins of Brahmaputra and Barak are combined with that of the Ganga, the composite Ganga-Brahmaputra basin becomes the single largest drainage area in the world, draining the waters of almost entire north India, into the Bay of Bengal via the Ganga, Brahmaputra and Meghana rivers through the Sundarban delta, which again is the largest delta in the world.

Figure 3

Ganga basin with its main tributaries
(without regard to international boundary).

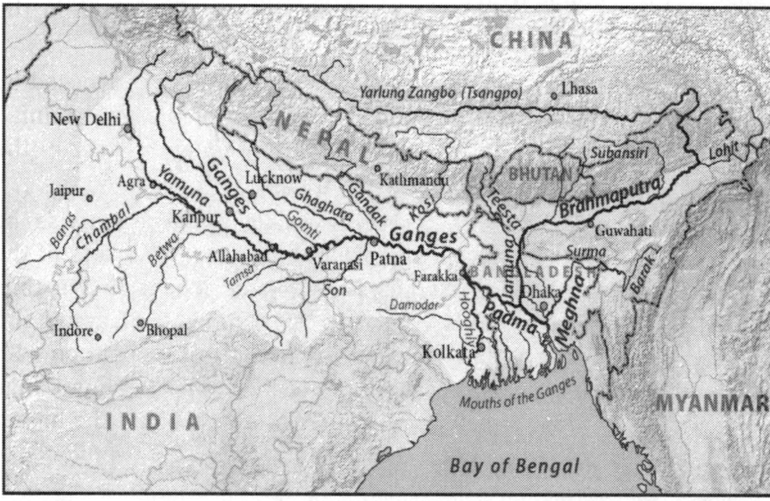

Source: Bluebird Electric Systems: http://www.bluebird-electric.net/oceanography/

Amongst India's mountains, the Himalayas are regarded as (geologically) young and are still rising. The rivers started to flow down the valley slopes as the Himalayas rose. The flow was highly turbulent and was able to mechanically erode the rocks and cut its way through them. Thus, some of the deepest gorges are found in the Himalayas, with the river flowing in the valleys. The mountain ranges in the north-eastern India are the youngest amongst the Himalayan mountain ranges; the region is, therefore, characterized by very narrow valleys and shallow alluvial plains in comparison with the Ganga alluvial plain.

As the rivers descended from their heights in the Himalayas, their initial north-south dispensation in the western Ganga (Gangetic) plain was forced to assume easterly direction by the Aravalli heights. They were, thus, obliged to follow the contour of the Tethys Sea bed and on which they deposited the alluvium which they carried in their waters. Similarly, the eastward flow of Brahmaputra was obstructed by the north-eastern Himalayan ranges and the river took a southward turn to enter India and to finally confluence with the Ganga.

Canals, Dams and Barrages from and on Ganga

Except for its flow and that of its tributaries in the Himalayas, The Ganga flows in plains with a hardly perceptible gradient. Its terrain is not suitable for the building up of dams but eminently suitable for the canal system. Much of its water was diverted away through the Upper Ganga Canal built in 1854 near Hardwar to irrigate the surrounding areas.

A multipurpose dam on Bhagirathi river which is the main tributary of Ganga in the Himalayas has been built and its first phase was completed in 2006. The dam creates a reservoir with a capacity of 4 cu. km of water. The dam is located in a major fault zone in the Central Himalayas. Other than concerns for the fragile ecosystem of the Himalayan foothills, there are concerns about the geological stability of the region. Close to the borders with Bangladesh, a barrage has been built on river Ganga in 1975.

In so far as tributaries of Ganga in the plains are concerned, Bansagar Dam has been built on Sone river besides the famous barrage at Dehri-on Sone. The largely monotonous relief of the plains is broken in its south-west portion centering on the Chambal river on which successive dams have been built. The Chambal basin has been notorious for its ravines or gullied badlands for harbouring violent gangs of dacoits.

Gully and ravines are two extreme forms of soil erosion. Gully formation depends on the runoff producing characteristics and gradient of the watershed and on the nature of soil. The depth of the gullies could vary between less than a metre to greater than 5 m. Ravines, on the other hand, are more severe form of soil erosion, resulting in the formation of complex network of gullies. In ravine lands, the drainage land has multi-directional slopes, resulting in the formation of a labyrinth like network of gullies.

Himalayas and Tibet Plateau: Store House of Global Distribution of Water

While Himalayas are the source of many Indian rivers and their tributaries, the Tibet Plateau is the single largest source of international rivers in the world for the supply of fresh water. These rivers are: Brahmaputra, Sindhu, Mekong, Yangtse, the Yellow river and Salween river. The territories of these rivers are facing the problem of supplying fresh water to their burgeoning populations. Natural rain water is no longer able to meet their demands of fresh water. Nepal and Bhutan, who have rich resources of water, have politico-economic compulsions to treat water as a tradable commodity, either directly or in the form of hydroelectricity but some others – Myanmar, Thailand, Laos, Cambodia and Vietnam – are helplessly watching the deteriorating water availability in their part of the international rivers. China, which is building multiple mega dams on all the major rivers running off the Tibetan plateau, is likely to emerge as the ultimate controller of water for nearly 40 per cent of the world's population. The net effect of the tremendous dam building activity in Tibet, or in the Himalayan region could be disastrous, because any reduction of water in their part of the international rivers could dry up their farmlands.

Many Himalaya geologists are apprehensive of seismic activity being triggered by the hydrostatic and gravitational pressure exerted by the mega reservoirs on the strata. This phenomenon is known as 'reservoir-induced seismicity'. The risks were highlighted by the severe damage caused to the 110-MW Rasuwagadhi Dam in Nepal by April 2015 quake of 7.8 magnitudes. Fortunately, the dam which has a capacity of 1.5 cu. km capacity was empty. Where the aim of the dams is primarily to generate hydroelectricity, the politico-economists are assuaging the apprehensions by generating electricity through what is called the 'runoff river' generating plants which either do not require any storage of water at all or require only a small or limited amount of storage in what is called the pond or pondage storage.

Indian geologists are apprehensive that as the Indian plate is pushing under the Tibetan Plateau at roughly 1.8 m per century, a great earthquake of magnitude of 8 or higher is likely to occur sooner or later, the severity of which will be more along the geological 'fault lines'. There is a strong possibility that some of the mega dams/reservoirs are located along them. The future of the world's most famous mountain range could be endangered by a vast dam-building project, as a risky regional race for water resources takes place in Asia.

The recent earthquake in Manipur, measuring 6.7 on the Richter scale is a serious wake-up call in this direction. Seismologists across the world have been expecting a great earthquake (> 8 magnitude) along the 400 km fault line in the Himalayas. The 2015 Nepal Earthquake of 7.8 magnitude fell just short of the mark and could be regarded as a prelude to a greater disaster, presently in making.

No doubt! Water is our necessity. But let us not repeat the story of the goose that laid a golden egg every day for its poor owner, who sold the egg to become rich. He soon became greedy and thinking that there must be storehouse of golden eggs inside the goose and that he should get them all in a day, he killed the goose. He lost not only the supply of a golden egg a day but also the goose!

The Western Ghats Water Shed

In comparison with the Himalayas, the Western Ghats are, geologically, older. The rivers on the plateau came only after the Ghats and the plateau had formed and the geological strata stabilized. When the rains came on the rocky mass of the plateau, water had to pave its way in the direction of the natural topographical slope of the stable surface strata, unlike the scenario in the Himalayas. Therefore, neither could they form deep gorges in the Western Ghats nor could a deep and comprehensively large plain of alluvium form. In contrast with the northern plains of the Ganga basin, those of the southern rivers are neither as deep, nor are they inter-connected.

The Aravalli ranges, Ramgarh hills, the Deccan and Chota Nagpur Plateaus and the Satpura and Vindhyachal mountains are also geologically older than the Himalayas. They have played a decisive role in defining not only the direction of flow of the rivers of northern India but also in determining the other water shed of the country, besides the Himalayas, i.e. the Central Indian Dome Shaped Plateau. The Western Ghats are as important for the river systems of southern India as the Himalayas are for the northern India.

Godavari River Basin

The 1,465 km long river rises in the Western Ghats water shed at Trimbakeshwar (near Nasik) on the eastern side of the Western Ghats, about 80 km away from the Arabian sea but flows south-east across Maharashtra (48.6% of total river basin of 3,12,812 km^2), Telangana and Andhra Pradesh (10.9%), Madhya Pradesh (10%), Odisha (5.7%) and Karnataka (1.4%) and empties into the Bay of Bengal forming a very fertile and heavily populated delta. Only Ganga and Indus (within Indian borders) rivers have a drainage basin larger than it.

A bulk of its waters, however, comes from the rivers originating from the third water shed in central India; prominent amongst these

Figure 4

Godavari river and its tributaries

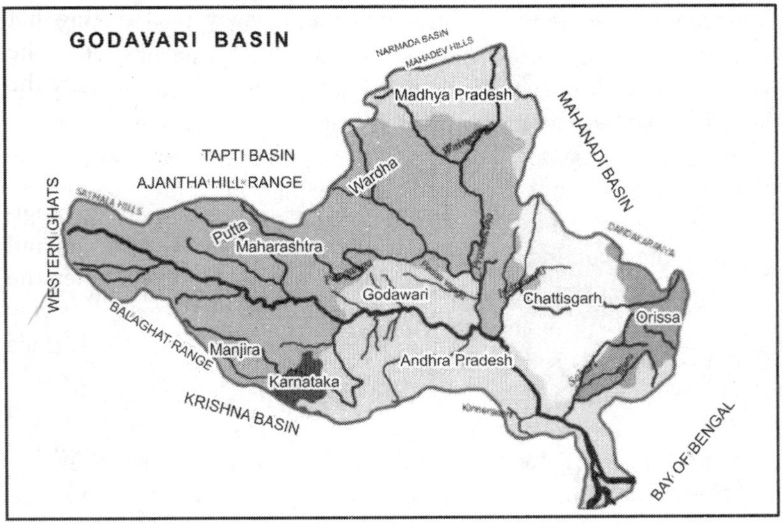

Source: Sakti, Database on Godavari Basin.

rivers are Poorna, Manjira, Pranhita, Indravati and Sabari; the last three which join in the lower reaches of the basin, carry three times more water compared to main Godavari. Figure 4 shows the river basin of Godavari. The river is often referred to as the Vridh or Budhee (Old) Ganga or the Dakshin (South Ganga). Inherent with the traditional nomenclature for the river is the geological fact that river Ganga is younger than Godavari river. The river has been mentioned in Veda and continues to harbour and nourish rich cultural heritages. Although the overall direction of slope of the Deccan Plateau is from west to east, it is the local topography that determines the direction of flow of the river; it is only after meandering in the locally available direction of topographical slope that the rivers are able to finally meet the Bay of Bengal, through the break in the Eastern Ghats. Though the heights of the Eastern Ghats is much lower than of the Western Ghats, their breaks along their longitudinal coastal spread play a decisive role in letting the rivers of the peninsular plateau meet the sea.

The Godavari basin is bounded on all sides by tablelands and hills and the entire drainage basin of the river system comprises rolling and undulating topography. This is eminently suitable for the formation of picturesque 'waterfalls', construction of dams and generation of hydro electricity. Besides, the basin is rich in deposits of coal and iron ore. As the river approaches the Bay of Bengal, it splits into two main distributaries, namely, Gautami and Vasishta. The delta consists of river borne alluvium. It is very fertile, densely populated zone (almost twice of the national average) and sees intense economic activity. The Godavari Basin has the highest number of coal-based thermal power plants which discharge high alkalinity water from their ash dumps into the rivers which seems to have seriously polluted the river and ground waters.

In the past few decades, the river system has been heavily tapped for its water through construction of barrages and dams. The Godavari basin has the maximum number of dams than any other river basin of India. It has the distinction of having the first barrage built at Dowleshwaram in 1862. The fallouts of large scale dam construction are (i) decreased water availability in river channels, and (ii) the arrest of much of the river sediments – silt, detritus and other matter – in the dams.

Whilst excessive damming of the river might be responsible for frequent drying up of the river bed in lean months of water flow, another and equally potent factor could be the irrigation of sugarcane farms by excessive drawing out of water from deep bore wells in Maharashtra, as a consequence of which the river begins to feed ground water resources; instead of the ground water contributing to the base flow of the river.

In the year 2015, the Godavari river was linked to the water deficit Krishna river by commissioning the Polavaram right bank canal with the help of Pattiseema lift scheme to augment water availability to the Prakasam Barrage located in Andhra Pradesh.

Krishna River Basin

Krishna, another important river, rises from an altitude of 1,337 m in the Sahyadri ranges at a place in Satara district just north of Mahabaleshwar, barely 64 km away from the Arabian Sea. Its basin extends over the states of Maharashtra, Karnataka, Telengana and Andhra Pradesh, amounting to a total of 2,58,948 sq. km, which is nearly 8 per cent of the total area of the country. The Krishna basin is bounded by the Balaghat range on the north, Western Ghats on the west, and Eastern Ghats on the east. Its principal tributaries joining

Figure 5

Krishna river basin

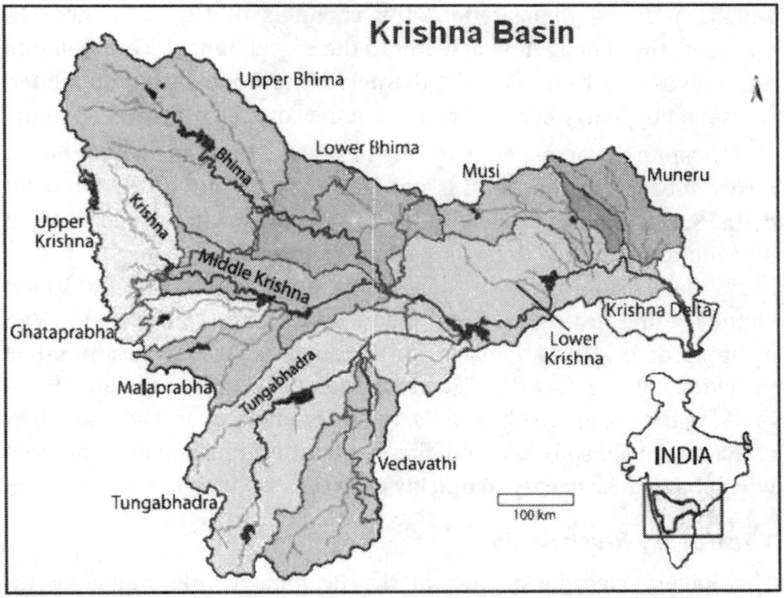

Source: National Institute of Hydrology, IWMI Report, Closing of Krishna Basin: Irrigation, Sreamflow Depletion and Macroscopic Hydrology.

from the right are: Malprabha and Tungbhadra and from the left are Bhima, Musi, and Munneru.

The per capita water availability in its basin (Figure 5) is less than the crucially important figure of 1,000 cum/person/annum to sustain a healthy life. Some parts of the basin, especially the Rayalaseema area of Andhra Pradesh, Bellary, Raichur, Dharwar, Chitradurga, Belgaum and Bijapur districts of Karnataka and Pune, Sholapur, Osmanabad and Ahmedanagar districts of Maharashtra are drought-prone. In order to cover this gap from the water available in the Godavari river, Godavari and Krishna rivers have been successfully inter-connected in September 2015.

Major part of the basin is covered with highly productive agricultural land. The average rainfall in the basin is 784 mm, mostly through the south-west monsoon in the mid-June to mid-October period. Most of the basin comprises rolling and undulating land mass. Unable to spread sideways during heavy downpour, the river flows deep, often reaching depths greater than 23 m with highly turbulent flow which results in heavy land erosion. The river carries a heavy sediment load which on being deposited in the channels of the distributaries, obstructs the speedy flow of water to the Bay of Bengal. This situation often gives rise to flooding in the delta. The flood situation is often worsened by heavy cyclonic rainfall in the coastal regions.

Despite the arrest of its waters in several dams and barrages, the heavy silt load in the Delta is unlike the behaviour of the Godavari delta. Some experts believe that the water carrying capacity of its distributaries in the delta is getting reduced as a consequence of heavy deposition of silt in their upstream channels. One important factor leading to the heavy silt load seems to be the loss of the 'soil gripping or grabbing' capacity of the surface through loss of cover provided by the forests and grasslands. The important soil types found in the basin are black soils, red soils, laterites and lateritic soils, alluvium, mixed soils, red and black soils and saline and alkaline soils, which have the capacity to quickly grab the rain water.

The Kavery River Basin

The Kavery river originates in the Brahmagiri Hill Range of the Western Ghats and drains a total area of about 81,155 km^2 in the states of Karnataka, Tamil Nadu and Kerala, which is about 2.7 per cent of the total Indian area. The basin receives rainfall mainly from the south-west monsoon and partially from north-east monsoon in Karnataka. But its basin in Tamil Nadu receives good water flows from the north-east monsoon. Physiographically, the basin can be divided into three parts – the Western Ghats, the Plateau of Mysore and the Delta. The delta area of alluvial soil is the most fertile tract in the basin. The other principal soil types found in the basin are black soils, red soils, lateritic soils, forest soils and mixed soils. Red soils occupy large areas in the basin. These types of soil rapidly absorb the rain water. Therefore, the runoff does not drain off quickly, whereby floods do not ordinarily occur in the basin.

The water deficient Kavery basin of the country has the distinction of holding several 'firsts' in the country in the sphere of management of natural water resources. First, several centuries ago, the first ever 'anicut' in the world was built on river Kavery by a King of Chola dynasty in the 1st century CE to take its water for irrigation to far flung areas; the word 'anicut' itself is from the Tamil language and entered the English language. Second, over a century and half ago, the enormous amounts of surplus water available in a river, originating from the Western Ghats and emptying into the Arabian Sea, was diverted into the east flowing river Kavery; and third, in more recent times, Tamil Nadu was the first state to make harvesting of roof top rain water obligatory.

The first hydroelectric power plant in the country was built as early as 1902 at Shivasamudram where the river falls through a height of about 91 m. As the river takes a southerly turn to enter Tamil Nadu, it enters the Mettur reservoir which was built in 1934. Nearly 100 dams have been built in the Kaveri basin to meet the requirements of irrigation and hydro-electricity is generated from a few of them.

Rivers From the Central Indian Dome-Shaped Plateau

Rivers Narmada and Damodar: Typical Examples of Beneficiaries of Rain Water Stored in Rocky Strata Faults

The sources of the Narmada River and the Sone are barely a kilometre away from each other in the Amarkantak Hills, close to the border between the Chhattisgarh and Madhya Pradesh states. In their runs thereafter the commonality between them ends. Narmada flows almost westwards and Sone northwards in geologically different types of terrains. The former is perennial and the latter, a typically monsoon river.

Narmada is a river in central India in Indian sub-continent. It forms the traditional boundary between north India and south India, and is 1,289 km long. It is one of only three major rivers in peninsular India that run from east to west, along with the Tapti and the Mahi rivers. Modern geo-scientists in India regard Narmada valley as a suture/mobile belt zone along which the southern Dharwar block (also called southern peninsular block) welded with its northern Bundelkhand block many millions of years ago, through tectonic movements. The suture/mobile zone is spread east to west with two well defined faults – the North Narmada Fault and South Narmada Fault. As was stated earlier, a 'suture' zone is created by the joining together of two, geologically separate terrains along a major fault zone. The suture is often seen on the surface relief in the form of a mountain range; in the present case, the mountain range is that of Satpura. The term is borrowed from surgery where it is used in the context of sewing together of two separate pieces of skull.

Narmada winds its course through the heavily faulted Mandla hills of magnesium dolomite, which form the head of the Satpura range; then near Jabalpur jumps into the Narmada valley between the Satpura and the Vindhyan ranges and persues a direct westerly course to the Gulf of Khambat (Gulf of Cambay). Beyond this point up to its meeting the Arabian Sea, the Narmada enters three narrow valleys between the Vindhya scarps in the north and the Satpura range in the South. Narmada is joined by many of its important tributaries from the south which bring the waters of the northern slopes of the Satpura Hills. Among them are: the Sher, the Shakkar, the Dudhi, the Tawa (biggest tributary) and the Ganjal. The Hiran, the Barna, the Choral, the Karam and the Lohar are the important tributaries joining from the north. Waters from the vast hinterlands make the river not only perennial but one that sustains several dams, besides the giant Sardar Sarovar, near Ahmedabad and is now expected to play an important role in the Kalpsar Project which seeks to create a reservoir of fresh water in the sea. Narmada waters have brought about quantum jumps in the agricultural and horticultural produce of Madhya Pradesh and Gujarat, besides meeting domestic and industrial demands.

Sone river originates very close to the source of Narmada river and flows north-northwest through Madhya Pradesh and then runs almost parallel to the Kaimur Hills of the Vindhyan Range and flows through Uttar Pradesh, Jharkhand and Bihar to become a tributary of the Ganga, near Patna. The river initially has a steep gradient (35–55 cm per km) with quick runoff and ephemeral regimes and becomes a roaring river with rain waters from its catchment areas but soon becomes a wide and shallow river. Its channel is as wide as 5 km at Dehri on Sone. After heavy rain in the hills even this wide bed cannot carry the waters of the Sone and disastrous floods in Shahabad, Gaya, and Patna are not uncommon. In summer, it nearly dries up over vast stretches with disconnected water pools with sand in between. In the past the Sone was notorious for changing its course. In modern times this tendency has been checked with Anicut at Dehri, and now more so with the Indrapuri Barrage, constructed in 1968 and the Bansagar Dam in Madhya Pradesh in 2008.

The Tapti River System

Not far away southwards from the Narmada, the river Tapti, also flows from east to west, almost parallel to Narmada but does not carry as much water as Narmada. It originates from a place called Multai in Betul district of Madhya Pradesh in Satpura Mountain range of Central Indian Dome Shaped Plateau. It is one of the major rivers of peninsular India with the length of around 724 km. The Tapi is the second largest westward flowing river that drains the rain water of the highly fractured and fissured rocky strata of the rocky Deccan Plateau through a very large number of tributary rivers flowing effectively in the south-north direction. The elevation of the origin of Tapi is about 650 m and its basin is essentially a highland of low rolling hills on the foot of mountains or hills. In geological terminology, it is called a Piedmont Plain. Its basin is a basaltic landscape with major physiographic units of plateau lands, escarpments, hills, piedmont plains, and valley plains.

The annual rainfall for the upper, middle, and lower Tapi basins for an average year is 935 mm, 631 mm, and 1,042 mm respectively. The Ukai Dam, constructed across the Tapi (or *Tapti River*) in 1972 is meant for irrigation, power generation and flood control. Having a catchment area of about 62,255 km^2 and a water spread of about 52,000 hectares, its capacity is almost same as that of the *Bhakra Nangal Dam*. It is known as the Vallabh Sagar.

The river, thus, gets an abundant flow of water which seeps out of the outcrops of porous or permeable rocks of Satpura ranges on the steep side of the valley. Narmada is able to feed not only the huge Sardar Sarovar in Gujarat, but several other dam-reservoirs in Madhya Pradesh as well. Narmada waters are feeding the Shipra river, on the banks of which is situated the holy city of Ujjain. The per capita/annum water availability in Narmada basin is 2,253 cu. m whereas that of Tapi, which almost runs parallel to Narmada, south of the Satpura mountains, is less than a third of this value, i.e. 731 cum.

The Damodar River System

Other important rivers in northern part of Dharwar Block – Damodar and Mahanadi – are also a beneficiary of water in the cracks and fissures of igneous rocks. Damodar river originates at a place called Chandwa in Jharkhand state in the Chota Nagpur Plateau. The Chota Nagpur plateau in its entirety lies between the basins of the Ganga and Sone Rivers in the north and the Mahanadi river in the south. Through its course, from west to east, it runs through the coal-bearing, faulted Damodar river valley. There are a number of coal seams in the valley and disposal of mine water has always been an issue. In order to illustrate the close relationship between faults and water habitats, we are citing the sad example of the Chasnala (near Dhanbad) coal mine disaster that occurred in December 1975. An explosion triggered the caving in of the roof in a section of the coal mine, the miners were trapped in under a mountain of debris and 372 mine workers were drowned when water surged in the area.

Numerous streams have dissected the uplands into a peneplain (a peneplain is geologically described as the flat land created by the denuding action of flowing waters on an upstream upland). The rocky strata of the plateau are heavily fragmented and highly dissected, resulting in the formation of many precipitous escarpments. There is a variety of rock formations. The metamorphosed granitic rocks are associated with porous lateritic rocks with sand and gravel which are good hosts of water. The area is drained by several streams and rivers which flow in all directions in highly faulted valleys. Damodar river originates and flows in a valley and receives the waters of many tributaries. This means that tributaries of Damodar river drain the waters of the monsoon rains of the entire sub-plateau into the river within a short time. As a consequence Damodar river swells in the rainy season and causes devastating floods, as if in rage and fury. The river was known as the 'Sorrow of Bengal', until tamed by several dams were built by the Damodar Valley Corporation in the post-1950 period. There are many waterfalls where rivers coming from the uplands of the plateau descend precipitously and enter areas of significantly lower heights.

The Chambal River Basin

Chambal originates near Mhow (near Indore, MP), almost 300 km north-west of Amarkantak, at an elevation of 853 m where there is an extensive spread of 30 m thick bed of lateritic minerals and lateritic soil which are known for their capacity to hold rain water. They provide shelter to the rain water which later continues to seep out of the rocks for the rest of the year, keeping the river alive. A number of dams have been built along its 625 km downhill flow towards Kota in Rajasthan to generate hydro-electricity, to increase the availability of potable water for domestic applications and to promote irrigation based agriculture and horticulture. The region has earned name for its long grained Basmati rice. No doubt, all this has improved the economy of the region. It should, however, be a matter of concern for all to debate if this is the best way to use the water for cultivating water guzzling paddy crop in a traditionally parched land, or to use it more effectively by cultivating crops that do not require intensive irrigation, such as the *arhar* or *toor daal*. It is gratifying that, through the efforts of agricultural scientists, this area is now becoming nationally known for the production of pulses.

The Mahanadi River System

Mahanadi is a river made by the confluence of several monsoon fed streams. It is one of the major rivers of the country and it ranks second to Godavari in its water potential and flooding capacity. It originates from a pool in Bastar Hills (Farsiya village in Dhamtari district of Chhattisgarh). The total length of the river is 851 km. It is fed by several tributaries such as Seonath, the Hasdeo, the Mand and the Ib rivers which join Mahanadi from the left, and the Ong, the Tel and the Jonk from the right. The basin extends over an area approximately 1,41,589 km^2, has a total length of 851 km and is bounded by the Central Indian hills in the north, by Eastern Ghats in the south and east and by the Maikala range on the west. Its basin is characterized by a tropical climate with average annual rainfall of 1,420 mm, with 90 per cent occurring during the south-west monsoon.

Mahanadi, was notorious for its devastating floods for much of recorded history. Thus, it was called 'the Sorrow of Orissa'. However the construction of the Hirakud Dam has greatly altered the situation. Today, a network of canals, barrages and check dams keep the river well in control.

After flowing through different geological formations of Eastern Ghats and adjacent areas, it joins the Bay of Bengal through distributaries to form the deltaic area. Geologists state that the channels of the river in the delta have changed the paths of their flow in the past, much in the same manner as the river as the river Kosi has done in its megafan.

Ordinarily, water in monsoon inspired rivers flows in them from the surface with gushing velocity. The fast flow of surface water on excessively grazed fields and deforested areas causes excessive erosion of the soil. As the waters carry away with them particles of the eroded soil, they eventually find their way into the Mahanadi. The river carries the highest silt load amongst the Indian rivers, mostly coming from its tributaries from the upper catchment areas. If any large water reservoir is built to store the excess waters of the monsoons, the silt will naturally settle down in the reservoir. It is reported that the designed water holding capacity of the Hirakud dam on the Mahanadi, which was amongst the first few dams built in resurgent India after 1947 is now, rather prematurely, severely compromised on account of much heavier deposition of silt in the reservoir than was anticipated in the original design.

Experts are of the opinion that the best way to slow down the process of siltation is to undertake plantation and soil conservation methods, besides preventing excessive grazing of grasslands.

The Mahanadi Delta is formed by the network of three major rivers: Mahanadi, Brahmani and Baitarini. The delta covers a coastline of 200 km, which stretches from south near Chilika to north up to Dhamra river. The Delta covers an area of 9,500 sq. km which consists of denuded hills and erosion plains. The depressions contain water bodies. The central part of the delta is distinct for its extensive plains and paleochannels (a palaeochannel is a remnant of an inactive river or stream channel that has been either filled or buried by younger sediments which can be either unconsolidated, semi-consolidated, consolidated. The satellite imagery of such paleochannels has been shown for the Kosi river megafan in an earlier chapter). The coastal parts of the delta contain lakes, creeks, swamps, beaches, tidal flats and mangroves. The delta contains critical conservation areas (in particular several fresh water and coastal wetlands) and biodiversity hotspots, with a wide range of species including rare and threatened ones. The delta is the ecological and socio-economic hub of state of Orissa, it supports more than one-third of its population, 68 per cent of which are farmers.

Two of the Mahanadi Delta's highly critical conservation areas are Chilika Lake, the largest brackish water lagoon (a coastal body of shallow water); and Bhittarkanika, a mangrove forest with high species diversity; and Garimatha, habitat of Olive Ridley turtles and the largest rookery in the world.

Basin of Rivers not Originating from the Three Principal Water Sheds

The Sabarmati Basin

The Sabarmati river basin is amongst the small river basins of the country. The 371 km long river originates from the Dhebar Lake, near

Udaipur (Rajasthan) in the Aravalli Hills at an altitude of 762 m. Its basin is bounded by those of rivers Manas and Mahi on the east, of Luni on the north, the Rann of Kutch on the west and the Gulf of Khambat in the south.

The river owes its existence to the monsoon and is ephemeral in nature. The rainfall in its basin varies from arid and semi arid in its smaller northern part, to tropical wet climate in its southern catchment areas. The rainfall varies from a few mm in Saurashtra to over 1,000 mm in the southern part. This difference in the rainfall in its basin arises because of the Aravalli Hills which influence the direction of the monsoon winds. Though the per capita/per annum availability of water in the basin is very low (263 cum), a number of dams and weirs have been built. The river bed remains dry for most part of the year. Since the internationally known Sabarmati Ashram of Gandhiji is located on its banks near Ahmedabad, the glory of the river is sought to be restored by reviving the flow with the help of water from the Sardar Sarovar.

Since this area has experienced geological activities, the bed rocks are highly jointed and fractured (geologically) and at many places unconsolidated deposits of sand, kankar and gravel are found. These unconsolidated formations rest on consolidated rocks, and are home to good aquifers at depths varying between 30 to 150 m in most part of the basin.

Inland Drainage Systems

The Ladakh and Sambhar Water Sheds

Water sheds drain into rivers and ultimately into the ocean. There are a few rivers in the world that do not reach the sea but end up in creating a terminal lake or a water body, scientifically called an endorheic 'waterbody' and the basin called an endorheic basin. The Aral Sea in Central Asia is a good example of closed drainage basin. This means that streams flow in but no water goes out of the water body/lake other than by evaporation or infiltration/seepage. In other words, if the river water is carrying dissolved salts or other pollutants – farm nutrients and pesticides, pathogenic bacteria, etc. – they get continuously concentrated in the water of the endorheic lakes. Some, like the Dead Sea, are so salty they cannot support life, but others may support very unique ecosystems. The difference in the ecosystems of fresh water and saline lakes is largely because of the difference in the concentration of dissolved oxygen in them; the greater the concentration of sodium chloride and other salts, the less soluble is oxygen. There are other factors too for their different ecological systems.

The geographical spread of the saline lakes/water bodies is large; they include the largest lake in the world, the Caspian Sea, the lowest lake; the Dead Sea; and many of the highest lakes, such as those in Tibet and on the Altiplano of South America. Formation of saline lakes requires compliance with certain geographical and geological conditions. These are (i) semiarid climate so that fast evaporation of lake water is ensured, (ii) a balance between the input of fresh water through rains and inflows through rivers, and output through evaporation (iii) the presence of endorheic drainage basins and (iv) geological strata bearing soluble minerals.

Notable examples of these conditions being met are in the Ladakh region and the Sambhar Salt Lake System in Rajasthan. The two regions are mutually independent and geographically unconnected and are also respectively independent of their neighbouring Sindhu (Indus) and the Ganga River systems.

A typical feature of Ladakh region is that a number of endorheic saline lakes exist here which serve as inland drainage receiver of the waters of streams in their areas. Among them is the Pangong Lake. The lake is partly in India and partly in China, spread in both the countries in the form of a narrow stream. Water of this lake is highly brackish with high content of subcarbonates of soda. The lake has its own unique ecosystem.

The brackish water of the lake has very low micro-vegetation, perhaps there is no aquatic life in the lake, except for some small crustaceans. But ducks and gulls around and on the lake surface can be seen. There are some species of shrubs and perennial herbs that grow in the marshes around the lake. The lake acts as an important breeding ground for a variety of birds including a number of migratory birds. During summer, the bar-headed goose and Brahmini ducks are commonly seen here. The region around the lake supports a number of species of wildlife including the kiang (an animal of a large race of the Asian wild ass with a thick furry coat, native to the Tibetan plateau) and the marmot (a heavily built, gregarious burrowing rodent of mountainous country area).

The Sambhar Lake receives water from a number of streams flowing into an endorheic basin with 5,700 sq. km catchment. Geographically, it is a land-locked river basin, located in Nagaur and Jaipur districts of Rajasthan, surrounded on all sides by the Aravali Hills. The level of water in the lake fluctuates between 50 cm in the dry season to about 3 m after the monsoon and is often referred to as saline wetland.

The Sambhar Lake is an important source of salt which is mostly produced by the evaporation process of brine. Sambhar Lake has been designated as a Ramsar site (recognized wetland of international importance) because it is a key wintering area for tens of thousands of flamingos and other birds that migrate from northern Asia. The specialized algae and bacteria growing in the lake support the lake ecology that, in turn, sustains the migrating waterfowl. There is other wildlife in the nearby forests, where Nilgai move freely along with deer and foxes.

The Luni river is the most important river amongst those in the internal drainage system. It begins near Ajmer in the Pushkar valley of the western Aravalli Range at an elevation of about 550m. At this point, the river is known as the Sagarmati. The river then flows in the south-west direction through the hills and plains of the Marwar region in Rajasthan. Its basin covers an area of 37,363 km^2, which includes Ajmer, Barmer, Jalore, Jodhpur, Nagaur, Pali, and Sirohi districts of Rajasthan and the Banaskantha and Patan districts of northern Gujarat. Maharaja Jaswant Singh of Jodhpur built the famed Jaswant Sagar Lake on it. The river serves as the primary source of irrigation in the region. It dissipates into the marshy lands of the Rann of Kutch, traversing a total of 495 km. The Luni is not saline until it reaches Balotra, where high salt content in the soil impacts the river. The river is also known as Lavanavati, which means 'salt river' in Sanskrit, due to the high content of salt in its waters.

Delta Regions of Rivers in Bay of Bengal

Water received as rainfall is consummated through storage in (i) surface reservoirs, (ii) in underground alluvial and unconsolidated rock aquifers, (iii) as soil moisture, (iv) as humidity or moisture in the atmosphere, and (v) the residual as run off to the seas through rivers. The runoff water from all the rivers of the world plays an important role in the eternal water cycle.

The place where the river meets the sea is called the 'mouth' of the river. In case the silt carried by the river is deposited at the mouth, and the sea is unable to wash it away, the resulting formation is called 'delta' and where the deposit is washed away by the tidal currents of the sea, the river mouth is called an 'estuary'. Indian rivers having their mouths in the Bay of Bengal form deltas but those having it in the Arabian Sea, form estuaries.

The Ganga, Brahmaputra and the Meghana rivers together form the world's largest delta, called Sunderban at their interface with the sea in Bay of Bengal, (Figure 6). The delta is spread over two countries –

Figure 6

A comprehensive photograph of the Sundarban Delta

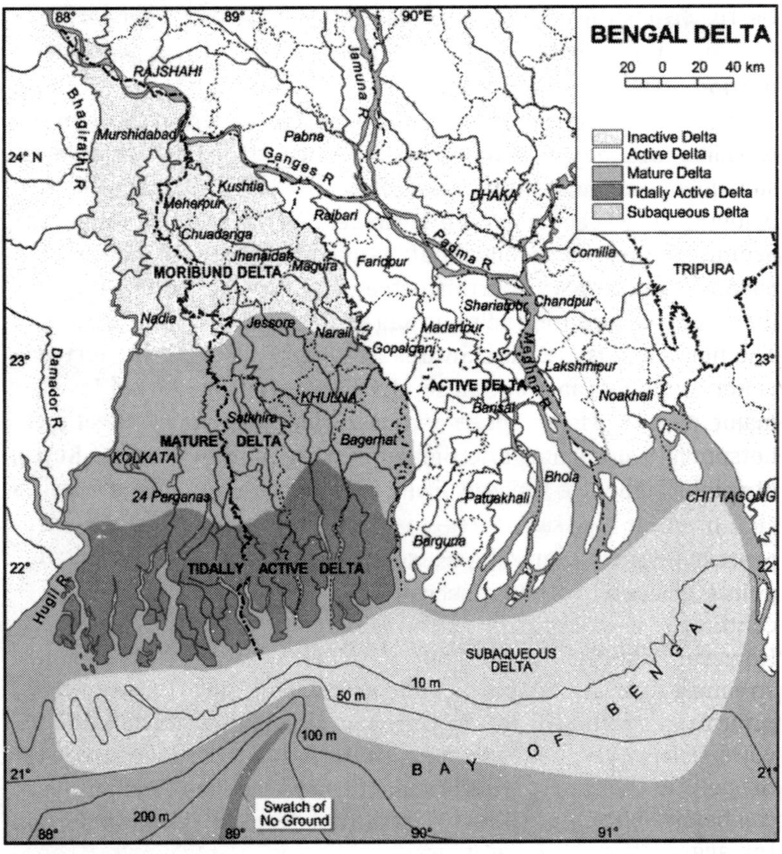

Source: All Bangladesh Map.

only 40 per cent of its 10,000 sq. km area is in India, the rest being in Bangladesh. The marked slowing down of the flow of the ageing rivers forces them to discard their silt load. The deltaic soil is, therefore, the naturally most fertile soil. However, as the waters move further towards the sea, their onward movement is obstructed by the ingress of sea water. This results in the settling down of their residual silt load and rather stagnant pools of water form, thereby creating swamps. The incoming river water becomes salty due to its admixing with the sea water. The situation is aggravated by the ingress of tidal waves from the sea. In order that the inflowing waters rapidly move towards the sea, despite difficulties created by sedimentation of silt, the river flow breaks down into a number of streams, now called distributaries, and the region acquires the physiognomy of the Greek letter, delta.

Deltas typically consist of three sections. The most landward portion is called the upper delta plain or 'stable' or 'moribund' delta. That portion of the delta which is affected by the ingress of sea water is called the tidally active zone of the delta and the third the subaqueous delta, which lies seaward of the shoreline and forms below sea level and marine processes dominate the system. Sediment-laden river flow that enters the ocean in well-defined channels loses its transporting power; the sediment is deposited and the river channels too end their journey. These are shown in figure 6. The figure also shows the Swatch of No-Ground which is an underwater canyon located 45 km south of Sundarbans. This is one of the major marine fishing zones and recently has been identified as a hotspot for cetacean diversity and the presence of at least four species of dolphins and two species of whales has been documented.

It may be appreciated that the deposition of silt in the delta is a process which is continuing from times immemorial. Two important points emerge from this discussion. First, the delta region should continually keep expanding as a result of continuous silt deposition. However, with the establishment of several dams and reservoirs up stream, a considerable portion of the silt load of the rivers is arrested prior to its reaching the delta region and gets deposited in the reservoirs. Usually, there is increase in both the area and land elevation in the delta as the supply of sediments (the alluvium) exceeds that which can be taken away by the sea. This is called the process of alluviation (or aggradation in geographical and geological terminology). As a consequence of the silt being arrested in the upstream dams the process of land building, or alluviation, may not be occurring; and if occurring, it may be less than the anticipated rise in sea level on account of melting of the ice of the polar regions due to global warming and increasing input water that has been drawn out from the ground. The net result would be submergence of these low lying lands of the deltas.

The Sundarban has been declared a world heritage site by the United Nations Educational, Scientific and Cultural Organization (UNESCO). It is our responsibility to ensure that the sanctity of its natural environmental ecology is maintained by all concerned. Besides being home to a variety of flora and fauna, including the famous Bengal tiger, the Sundarbans are home to over 85 varieties of mangroves. Mangrove is a unique tree which survives in salty sea water and helps to arrest the onslaught of sea waves on coastal land, hence check erosion of the land mass.

Mahanadi, Godavari, Krishna and Kaveri also cause the formation of deltas at their mouths with the Bay of Bengal. A study found that the suspended sediment load at the Godavari delta has reduced from 150.2 million tonnes during 1970–1979 to 57.2 million tonnes by 2000–2006, which translates into a three-fold decline in the past four decades. This decrease is attributed to the sediment load of the river getting deposited in the large reservoirs built upstream. The Godavari delta has, therefore, been categorized as having substantial risk of flooding with rising sea levels and destruction of mangrove forests.

Space imagery provides indications that the Kaveri delta may also not be growing now as the several reservoirs built upstream arrest the silt flowing in them along with the water.

Effect of Anthropogenic Activities on Water Bodies and Habitats

Effect of Global Warming on Himalayan Rivers

On account of the world-wide concern about global warming, the Himalayan glaciers have become the subject matter of global interest and study. Close monitoring of the movement of the glaciers by international space agencies, including by the Indian Space Research Organization, has revealed the startling truth that the glaciers are receding or retreating at a fast speed; whilst some have already vanished, many others in Nepal and Bhutan are in highly threatened state. It is a matter of deep concern that the Gangotri and the Yamunotri glaciers are receding at a fast rate and are presently in a highly threatened state. We have seen in an earlier chapter only one facet of the disastrous consequences of the shrinking of the glaciers – that of Glacial Lake Outburst Flow. The survival of all the Himalayan glaciers and restoration of their pristine glory is a matter of vital interest, not only for us but also for all South-Asian and South-East Asian countries, including China.

Global warming is also causing the melting of ice of the permafrost which had formed thousands of years ago under the ground in the very high altitude valleys of the Himalayas as well as in the Arctic region. As a consequence of the occurrence of this phenomenon in Tundra (Arctic Circle), huge quantities of carbon dioxide and methane gases are released into the atmosphere. These gases are formed in the permafrost layer through the putrefaction of the vegetative mass which was embedded in it but otherwise remained locked therein. Their release on melting of permafrost effectively supplements the continuously increasing concentration of green house gases in the atmosphere. Their presence in the atmosphere is regarded as the very cause of global warming.

The thawing of the ice in permafrost occurs slowly. As explained in an earlier chapter, the consequences of the phenomenon which occurs in the higher ranges of the Himalayas may happen catastrophically. They then inflict devastation and misery on the population living in the lows of the valleys through floods. Further, as the water which had hitherto remained arrested within the erstwhile permafrost bed is suddenly released, it gushes down the slopes with torrential velocity; pieces of rocks also roll down the slopes with it, causing immense havoc as they do so. That such events are now taking place with increasing frequency in Kashmir and elsewhere in the Himalayas is a matter of concern for all. The recent (2013) catastrophic disaster through flash floods at the Kedarnath shrine was one of its most cruel manifestations.

Apart from global warming, human activities, concerning mostly deforestation, have also caused the degeneration of the network of their root systems. Together with the top soil, the roots form sponge-like composites. As long as this sponge is organically alive, it not only holds rain water but also keeps the top soil, *in situ*. The destruction of this natural 'sponge', loosens the top soil which now flows down with water, and increases the silt and mud content of the river waters. Such events cause acute hydrological stress on the rivers and their ecosystems. Already, some of the rivers which were regarded perennial until a couple of decades back are experiencing severe reduction in the quantity of water in their flow, particularly in lean months. As a consequence, dry patches often appear randomly, at some time or the other during the year, somewhere along their flow path in some of the rivers.

In this connection, mention may be made of Yamuna and Gomati, which were until a few decades ago regarded as perennial, are no longer so. Their behaviour is becoming akin to that of the ephemeral rivers. Only the names of these two rivers are mentioned but they are merely symbolic of a more widespread occurrence.

Impact of Human Activities on the Water Bucket of the Deccan Plateau

We have seen that the assiduously carved geomorphology of the Western Ghats and the Deccan Plateau have created the 'transit stores' of rain water to provide time for the infiltration of rain water into the ground. Were human activities to disturb this naturally crafted 'transit store', it could seriously compromise with the amount of water infiltrating into the ground which will reduce the flow of water in the rivers. There is documented evidence to show that human activities through surface mining and deforestation on the Western Ghats have disturbed their ecology. When the highly porous surface beds of lateritic soil and the lateritic iron ores are, *en masse*, removed in the mining operations, the transit route for rain water to enter the ground water is destroyed. As a consequence, more water runs down the slopes of the Ghats to the Arabian Sea and correspondingly less water is available to infiltrate and replenish the ground water resources. It is a matter of common knowledge that the continuing drought since 2012–13, or so, has made a bad situation worse in the water-stressed regions of Karnataka and Maharashtra as underground water level has gone so down that it is beyond the reach of powerful bore pumps. As we have seen in the preceding chapter, there are other potent causes that have contributed to the development of this scenario.

Hydro-geological experts are now veering round the point of view of environmentalists that certain human activities that have been or are being executed without any regard for the protection of naturally created/formed store houses of water (both on land surface and underground), have acutely worsened the situation of water scarcity which is caused by the inadequate monsoon rainfall.

The Western Ghats are a national treasure of water resources for almost the entire southern India. Their virginity and sanctity must be preserved in order to provide water security to our population in parts of Maharashtra, Karnataka, Telangana, Andhra Pradesh and Tamil Nadu. Any activity that damages the physical and environmental or ecological integrity may be regarded as an act of treason.

Another consequence of the environmentally unfriendly human activities is its effect on making the monsoon behaviour rather erratic in the region. This has possibly caused a change in the behaviour of monsoon rains – which are at times exceedingly heavy, followed by long dry spells. Though the statistical average of rainfall may remain unchanged despite the changed pattern of rainfall, the behavioural consequences of erratic rainfall could be drastically different. The natural process of infiltration of water into the bowels of the earth is a time dependent phenomenon. But the sudden downpour of excessive rain does not give that time to percolate downwards. The excess water naturally 'runs off' to the rivers, thereby creating situations that favour the occurrence of flash floods, and finally winding its way to the sea through the river. The development of such situations is regretfully not merely confined to one region alone but is beginning to occur on wider scale across the country.

Effect of Human Activities on Inland Saline Lakes

The importance of endorheic salt lakes as an integral element in biospheric processes was until recently generally overlooked. Global attention on the vital role played by them in determining the local ecology was focused when the Aral Sea (a large saline lake in Central Asia) began to shrink as a consequence of the diversion of fresh water which used to feed the Aral Sea for agricultural purposes. The level of the lake fell, salinity rose, and vast expanses of the dry lake bed were exposed. As a result, 'the fishing industry collapsed, islands that had served as wildlife habitats became devoid of biological diversity and productivity fell, biota disappeared, large quantities of salt blew from the lake bed onto neighbouring lands, ground water salinity rose, and the local climate was altered. The effects on the local human population were catastrophic as well' (*Encyclopaedia Britannica*).

In India too, water bodies have begun to lose their pristine character as a consequence of human activities interfering with their systems. Most Indians are more familiar with the word 'sambhar' as a variety of natural, unrefined salt obtained from Rajasthan than with its source, i.e. the saline Sambhar Lake itself. That the fresh water reaching the Lake turns saline in the Lake is a wondrous act of Nature that has been going on unabated for centuries. As the water inflow in the lake is through fresh water rivers, which in itself does not carry any salt in solution, the source of the salt in the lake waters has always intrigued people. It is clear that the salt in the lake is not the result of concentration build up in the residual solvent (i.e. water) on account of progressive evaporation of inflow water. As the composition of the brine in Sambhar Lake is typically different from marine brine, the source of the salt in the Lake cannot be traced to the sea. It is relatively deficient in calcium and magnesium and has higher amounts of sodium salts as carbonates, sulphates and chlorides. Geologists believe that the salt originates from the water induced solution weathering of the strata of the catchment area of the Lake. Water infiltrating into the ground picks up the salt from the bed strata. The now salt-rich water rises through capillary action to the lake bottom surface. As the lake gets completely dehydrated during summer, an efflorescent crust of salts is formed on the surface. In chemistry, efflorescence is the migration of a salt to the surface of a porous material (lake bed in our case). The essential process comprises the dissolving of an internally-held salt in the bed strata. The water, with the salt now held in solution, migrates to the surface, then evaporates, leaving a crust of the salt on the surface. In other words, the infiltrating water is the invader to rob the strata of the salt physically present in the rock minerals of the bed strata. Salt production in the Sambhar Lake is, therefore totally dependent on the quantity of infiltrating water. We can state this fact by stating that more the infiltration in a wider area, the higher would be the salt production; along with this, the higher the atmospheric temperature, the more will be salt production.

In recent years, however, there is a decline in salt output. Experts attribute the decline to the catchment of the lake receiving less water due to human interference. During the last few years, there has been a steady decline in the water flow in the basin through construction of dams and diversion of water for agriculture. The Lake area is shrinking fast. The situation is worsened by salt manufactures themselves by drawing out underground brackish water through bore wells thus deepening the water table so deep that capillary action in the so-called fringe zone is unable to naturally 'pump' up the brine.

Further, the fall in the water level in the lake is straining the lake's ecosystem as well. Sambhar Lake used to attract thousands of flamingoes. But since the lake now remains dry for most part of the year and there are reduced quantities of algae available for them to eat, fewer flamingoes are visiting it. It is high time that the Indian society takes note of it and strictly enforces the steps recommended by appropriate authorities to ensure the sustainability of the lake and other water bodies.

Effect of Human Activities on Local Ponds

In post-independence India, the dire necessity of rapidly increasing food crop production and to provide settlement to vast exodus of population from what now are Pakistan and Bangladesh led to the adoption of some environment unfriendly acts in order to make additional lands available for agriculture and rehabilitation of displaced persons. These led to decreased ground water recharge. In particular these acts were:

- Rampant deforestation in the Terai regions at the low land foot beds of the Shivalik ranges and central India, and *en masse* eradication of the kans or *kush* grass (saccharum spontaneum), the long and deep rooted fine network of which formed a bio-sponge with soil, helped to retain soil moisture for a longer time and provided an underground pseudo barrier for the movement of moisture along the gradient of the strata; i.e. the bio-sponge functioned like an underground check dam. The net effect of kans grass eradication was considerably decreased ground water recharge and increased silt load of rivers. Kans is a perennial grass, growing up to 3 m in height, with spreading rhizomatous roots, kans grass quickly colonises exposed silt plains created each year by the retreating monsoon floods Kans grasslands are an important habitat for the Indian rhinoceros (*Rhinoceros unicornis*). In Nepal, kans grass is harvested to thatch roofs or fence vegetable gardens. It quickly colonizes other lands only to become an invasive species that takes over croplands and pasturelands. It is very effective in preventing soil erosion. Perhaps, because of its ecological importance, kans grass is used in many Hindu rituals.

- Switch over to irrigation from ground water, and consequent neglect of ponds accelerated their decline. In course of time, they became overgrown with weeds. The epidemic of installing bore pumps for water fanned the crisis.
- Gradual abandonment of pasture lands and livestock watering ponds. Countless ponds in the alluvial soils and many karst ponds popping up on the surface were filled up to meet this need in central India.
- The recharge of ground water declined and water table receded deeper in the ground. In areas where the bore well was withdrawing water from the aquifer formed in the cavernous rocky underground strata, many cavities/caverns became empty as a consequence of excessive pumping out of water. The situation was often aggravated by continued occurrence of natural drought.
- Legal and illegal sand mining is rampant on many river beds; the removal of permeable sand bed caused severe reduction of ground water recharge and increased the runoff water in the river; hence severe reduction in ground water recharge.

Effect of Human Activities on the Formation of Sinkholes in River Beds

The formation of sinkholes in river beds on account of human activity is a widespread natural phenomenon that occurs the world over. At many places in the world, rivers are flowing over strata where the underlying rocks are cavernous limestone, with water in the caverns making the strata mechanically stable to withstand the weight of the overburden. At other places, the strata may be hosting unconsolidated sand and gravel beds with high inter-gravel porosity below the water table. Both these geological situations lead to the formation of rich aquifers. Often excessive withdrawal of ground water from the 'Karst' and/or unconsolidated aquifers makes the karst caverns or the inter-gravel spaces respectively hollow. Thereby the overburden strata become mechanically unstable and susceptible to dramatic collapse.

Thus, a sinkhole had dramatically formed in the bed of river Chitravati at Guddumarri village in Anantpur district of Andhra Pradesh in late January 2015. The sinkhole had a diameter of about 8 m and was about 8 m deep. Textbook conditions for the formation of sinkhole were present at the time of the incident: namely, (i) presence of cavernous limestone bed at a depth of about 80 m below the river bed, (ii) removal of the highly porous river bed through rampant sand mining, (iii) excessive pumping out of ground water to irrigate the fruit orchards and farms. The village is in perennially drought ridden area and there has been lack of good rain for the past few years. Geological investigations of the incident revealed that the cavities in the limestone bed were devoid of any water due to heavy pumping out of water through bore wells deeper than about 250 m.

The second most recent example comes from Mexico, where desert and semi-desert conditions prevail over two-third of the country. Ground water is the principal source of water in Mexico. Heavy withdrawals of ground water for domestic, irrigation and industrial needs since the 1970s have led to ever deepening of the water table and increasing consequent threats of collapse of underground dry caverns. A giant fissure (30 m x 20 m) opened up dramatically on a March 2016 night in the Atoyac river and water of the river disappeared overnight into it.

Sinkholes can also form when water drainage patterns are changed and new water diversion systems are developed by human activities. The most dramatic example comes from the Dead Sea region. The surface and shores of the Dead Sea, bordered by Jordan in the east and Israel and the West Bank in the west, are about 430 m below the mean sea level. Geologically, beneath the clay-like surface layers, there are cavernous spaces which, until a few years ago, were filled with water and provided stability to the over-burden. In recent years, the waters of the Jordan river, principal tributary to the Sea, have been diverted with the consequence that level of Dead Sea has been dropping very rapidly – a metre a year as input of fresh water has practically dwindled but evaporation goes on unabated. It has dropped by over 30 m in the last few decades. The shoreline of the Dead Sea has become notorious for the abrupt formation of 'sinkholes' just anywhere. Today, more than 3,000 sinkholes are known to exist along its banks, some as deep as 25–30 m. Experts believe that the formation of sinkholes is inevitable because with the receding of the water table, the subterranean cavernous spaces are progressively drying up. Consequently, their surface layers are becoming unstable and dramatically collapse into the emptied spaces, creating chasms along the banks of the sea. Further, the caving in of surface layers could also be triggered by heavy rains or earthquake as well. Experts believe that there is only one way to redeem the situation and that is by restoring the diverted water of the Jordan river back to the Dead Sea.

Natural sinkholes influence the course of flow surface streams, particularly where the underground geological strata host interconnected karst formations. In such regions, natural drainage of rain water may occur intermittently through both surface streams and karst subterfuge of underground streams. There are several examples of such situations the world over.

The Alapaha river in the United States of America is an example of such a river which has two large sinkholes in the course of its flow. During periods of low flow volume in the river, it disappears underground through the first sink hole to become a subterranean river for some distance. Reappearing on surface, it again disappears into another sinkhole downstream. At the latter point, the river disappears underground almost completely, leaving a dry bank for much of the remainder of its course. Another river in the USA derives its name from this phenomenon and is called 'The Lost River' simply because in its course of about 140 km run, it flows at least 37 km completely underground disappearing into, and coming out, a series of underground caverns. 'The Lost River' begins as a normal river, but soon after rising, the river flows over and into a limestone bed (karst) for several miles until the stream bed turns dry; the water is absorbed into the limestone and sinks beneath the surface to hidden caverns. Geologists have estimated that there may be more than 400 sinkholes in 1 sq. km surface area.

Europe's mighty river, Danube, springs up as two small streams in Black Forest Mountains of Germany and flows through a strata which partially consists of limestone with faults and 'Karst' crevices and caverns. The river waters enter these underground caverns and sometimes cause complete disappearance of the surface flow. The number of days its surface flow completely disappears has been rising since it was first recorded in the year 1874. The fact that Danube waters enter the underground water passages through sinkholes and appear elsewhere as springs was proven in the year 1877. Geologists poured three chemicals in the river just ahead of a sinkhole; they found that these materials came out through a spring, 60 hours later, and proving thereby that the spring was connected with the river flow.

The Ramsar Convention for Conservation of Wetlands

Enlightened scientists, ably supported by geo-politicians, began to express concern at the rapid degradation and disappearance of wetlands in all parts of the world. The Ramsar Convention on Wetlands was developed under the United Nations as a means to call international attention to the rate at which wetland habitats were disappearing, in part due to a lack of understanding of their important functions, values, goods and services. The Convention on Wetlands, signed in Ramsar, Iran, in 1971, is an inter-governmental treaty which provides the framework for national action and international cooperation for the conservation and wise use of wetlands and their resources. There are presently 158 contracting parties to the Convention, with 1758 wetland sites, totalling 161 million hectares, designated for inclusion in the Ramsar List of Wetlands of International Importance. Ramsar Convention is the only global environment treaty dealing with a particular ecosystem – that of wetlands.

The Ramsar Convention on Wetlands defines wetlands as: 'areas of marsh, fen (an inland area of low lying marshy land, now often drained and cultivated because of its nutrients-rich soil), peatland (an area of land consisting mainly of peat bogs) or water, whether natural or artificial, permanent or temporary, with water that is static or flowing, fresh, brackish or salt, including areas of marine water the depth of which at low tide does not exceed six metres'. The natural wetlands occur in areas where the water table is at or near the surface of the land or where the land is covered with water. The Convention uses a broad definition of wetlands. It includes all lakes and rivers, underground aquifers, swamps and marshes, wet grasslands, peatlands, oases, estuaries, deltas and tidal flats, mangroves and other coastal areas, coral reefs, and all human-made sites such as fish ponds, rice paddies, reservoirs and salt pans.

All wetlands provide natural biodiversity and have their own distinct and characteristic ecology associated with plant, animal or bird life. They are the habitats of countless aquatic plants, from single cell organisms to vertebrates to flowering plants. In all cases, however, water is the primary factor controlling the ecology. The plants and animals have adapted themselves to changing water levels, dissolved salts and salinity, different contents of dissolved oxygen content and climatic conditions. The diversity of plant and animal life also depends upon the type of water source, as well as its size, depth, altitude and other local habitats. Aquatic plants are different from land plants though both produce their 'food' through photosynthetic process which requires carbon dioxide, water and sunlight; aquatic plants get water and carbon dioxide from their aquatic environment as well as sunlight because sunlight can pass through water, though in decreasing intensity at greater depths. So aquatic plants cannot grow at deeper depths where sunlight is not available.

The density of birds, local and migratory, regarded by experts as an accurate indication of the ecological health of a particular wetland. The other types of life include dragon flies, variety of amphibians, reptiles and mammals whose life cycle is connected with the water in the wetland.

Wetlands are among the most diverse and productive ecosystems. Despite their great importance in defining the socio-economic features of a geographical region, the wetlands continue to be degraded and converted to other uses. The Ramsar Convention aims to conserve and promote wise use of all wetlands through local and national actions and international cooperation for achieving sustainable development throughout the world. More than 25 wetlands have been identified in India under the Ramsar Convention for conservation and to ensure that the use of their waters does not lead to further deterioration. Their list is continually expanding. Some amongst them are the following:

1. The Bhittarkanika Mangroves wetland in the delta of Brahmani and Baitarni Rivers in Odisha. They are home to 55 of 58 of India's known mangrove species, largest population of saltwater crocodiles and include the nesting beach for olive ridley sea turtles.
2. Loktak Lake in North-East region of the country and famous for the 'phumdis' which is heterogeneous mass of floating islands consisting of mass of vegetation, soil, and organic matter at various stages of decomposition. They cover a substantial part of the lake area. The largest single mass of phumdi is in the south-eastern part of the lake, covering an area of 40 km^2. This mass constitutes the world's largest floating park, named Keibul Lamjao National Park. The park was formed to preserve the endangered Eld's deer subspecies, called Sangai in the Manipuri language. Phumdis are used by the local people for constructing their huts for fishing and other livelihood uses, and are inhabited by about 4,000 people.

3. Ashtamudi Wetland is the second largest estuarine system in Kerala and clam fishery began in 1981 which are largely exported to Vietnam, Thailand and Malaysia. It supports the livelihood of around 3,000 fishermen.
4. Chilka Lake is a brackish water lagoon in Odisha state at the mouth of the Daya river. It is the second largest lagoon in the world and the largest wintering ground for migratory birds on the Indian sub-continent and hosts over 160 species of birds; birds from as far as the Caspean Sea, Lake Baikal, Kirghiz steppes of Mongolia, many remote parts of Russia, and Ladakh and other Himalayan parts of India come here. The lagoon is also home to 14 types of raptors. Around 152 rare and endangered Irrawaddy dolphins have also been reported. Plus, the lagoon supports about 37 species of reptiles and amphibians. Microalgae, marine seaweeds, sea grasses, fishes and crabs also flourish in the brackish water of the Chilika Lagoon.
5. Keoladev National Forest, Rajasthan, is a man-made wetland. The reserve protects Bharatpur from frequent floods, provides grazing grounds for village cattle. The 29 km^2 wetland is a mosaic of dry grasslands, woodlands, woodland swamps, and wetlands. These diverse habitats are home to 366 bird species, 379 floral species, 50 species of fish, 13 species of snakes, 5 species of lizards, 7 amphibian species, 7 turtle species, and a variety of other invertebrates. Every year thousands of migratory waterfowl visit the park for wintering and breeding. The sanctuary is one of the richest bird areas in the world. The rare, and now probably extinct, Siberian cranes used to winter in this park.
6. Nal Sarovar, near Ahmedabad, is the largest wetland bird sanctuary of Gujarat, consisting primarily of lake and ambient marshes. It is visited by migratory birds in winter and spring. The lake attracts over 210 species of birds in the winter, and harbours a variety of plants and animals. Besides a few mammalian species, including the endangered wild ass and the black buck, its migratory bird population includes rosy pelicans, flamingoes, white storks, brahminy ducks (also called Ruddy Shelduck) and herons. Thousands of migratory *waterfowl* flock to this sanctuary just after the Indian monsoon season. The shallow area and ponds on the outer fringes of the lake attract the wading birds that feed in the shallow waters. It harbours over 250 species of wetland birds including purple moorhen, pelicans,

lesser and greater flamingos, white storks, four species of bitterns, crakes, grebes, brahminy ducks and herons.

7. Rudrasagar Lake, Tripura, is a natural sedimentation reservoir, which receives flow from three perennial streams. The sediments brought by the inflowing streams settle down to form the silt bed of the lake and the clear water discharges into the river Gumati. Annual rainfall of 2,500 mm is spread over the months of June to September. Lake water is fresh with a depth varying from 2 m to 9 m. Rudrasagar attracts a large number of waterfowl in winter. Among the rarer species recorded are the endangered Baer's Pochard and near-threatened Ferruginous Duck.
8. Hokera Wetlands, Kashmir, is a natural perennial wetland in the snow draped Pir Panjal ranges, about 10 km away from Srinagar. It is the only remaining site for the reedbeds of Kashmir and a number of other typical marshy vegetation. It is visited by migratory birds from Siberia, China, Central Asia and Northern Europe.
9. The Punjab Wetlands: Three of Punjab wetlands have been identified as the Ramsar sites. These are the Harike, the Kanjli and the Ropal Wetlands located respectively at the confluence of rivers Satluj and Vyas, on river Holy Kali Bein and on Satluj river in the lap of Shivalik foothills. They are among the large wetlands in northern India. These wetlands are important from the point of view of being aquatic ecosystems, bird sanctuaries or have sanctimonious socio-religious significance. The oldest canal system – the Sirhind Canal System – is based on the Kanjli Wetland. The wetlands are facing severe problems of siltation and loss of territory, inflow of pesticides, fertilizers through runoff from agricultural fields and sewage, invasion and growth of weeds, etc.

In addition, many wetlands are international systems lying across the boundaries of two or more countries, or are part of river basins that include more than one country. The health of these and other wetlands is dependent upon the quality and quantity of the trans-boundary water supply from rivers, streams, lakes, or underground aquifers. This requires framework for international cooperation toward mutual benefits.

The Concept of Micro-Level Water Sheds

Large water sheds may contain many smaller water sheds. Thus, the Himalayan water shed consists of many sub-water heads if we regard each individual river, originating from the Himalayas as an independent common outlet for all rivulets and streams draining into the river. Thus, if Ganga is regarded as the common outlet, its water shed area keeps progressively increasing as the river flows onwards. It rises from the Gangotri Glacier in the (Garhwal) Himalayas at an elevation of some 4,100 m above the sea level under the name of Bhagirathi. This main stream of the river flows through the Himalayas till another two streams – the Mandakini and the Alaknanda – join it at Dev Prayag, the point of confluence. Its water shed or catchment area of the water that enters its flow keeps ever increasing until it drains itself off in the Bay of Bengal.

India is drained by many rivers and their tributaries (and sub-tributaries); floods and droughts occur frequently. The concept of macro-sized water sheds (common outlets) has helped to plan location and size of dams and reservoirs along the paths of rivers. Whilst the country has immensely benefitted from these large reservoirs of water, it fails to take into account the local scenario of water availability (rainfall, local streams or replenishment of ground water resources) or involve the local populations in the water management activities. This resulted in the neglect of local water sheds and caused widespread lowering of the water table due to relentless withdrawal of ground water. To combat these situations and sustain the local ecology – flora and fauna – and agriculture, local geomorphology and local water sheds have to be taken into account. In order to do so, the experts decimated the large common outlets or the macro water sheds into smaller and smaller units till it comes down to micro levels of individual ponds or nullahs, including seasonal nullahs which get activated only during rains. It is almost the same as the country being divided into states and districts for better administration. The country has been thus divided into river basins, catchment area of rivers, sub-catchments, water shed, sub-water shed and micro water sheds. These delineations have been extensively used in the planning and development of surface water resources of the country.

Benefits from Micro-Level Water Management of Resources

The lowest, unconnected depressions in undulating land topography may give rise to the formation of a pond when the dimensions are small or a lake when the dimensions are large. Management of water resources at local level takes note of such features in a micro-basin. Several benefits accrue through this approach. Some are indicated below:

- Conserves rain water where it falls; every drop of rain water must be conserved.
- Draining out excess water and diverting it to storage ponds/*johads* and the like. Reduces floods and flash floods.
- Arrests soil erosion which otherwise would have been caused by rapidly flowing runoff rain water.
- Improves soil moisture content thereby ensuring vegetative cover on the surface. Reclaims vast tracks of eroded land.
- Involves participation of local people in the management of their natural water resources and brings people closer together specially if they took part in the restoration process together.
- It does not involve massive outlay of capital expenditure or massive displacement of populations from the area.

This approach has been vociferously advocated by Rajendra Singh – globally decorated as the 'waterman of India' – and his Alwar based (Rajasthan) Tarun Bharat Sangh (a non-governmental organization to improve the scenario of water availability, both surface and ground water in several drought affected regions of India – most noticeable success being the revival of long lost Arvari river in Alwar district of Rajasthan. Inspired by the success of their efforts, a farmer in a remote village in Dewas district (Madhya Pradesh) successfully made a 2 m deep pond in his own farmland to store rain water and used the water to irrigate his fields. This ushered a new life to his farmland. This small success story of 2004 had a ripple effect that has transformed the once-parched district in the state's Malwa region into a lush green area. Since 2005, about 6,000 farm ponds of sizes varying from a quarter to 5 ha have been dug. They have changed the agricultural landscape of Dewas where water crisis had affected crop cultivation. Water levels, which had dropped to 150 m below ground, are now up, in some areas at less than 15 m. Planning the methodology for the conservation of rain water on considerations of local topography allows freedom to adopt the most optimum technique from the available options: check dams, *johads, baoris*, ponds, ground water recharge.

A knowledge of the local topography, soil type (clayey, sandy or silty), rock formation, may help to maximise benefits from the newly constructed ponds. Where the objective of the pond is to store rain water exclusively for irrigation the pond may be paved with consolidated, unfired bricks/tile of clay to reduce percolation of water into the ground; where the object is to enhance ground water recharge, the pond may have sandy or silty bed.

Let there be no ambiguity about it. Bore wells have their own place in water management programmes, specially in regions where the sub-surface is non-alluvial. Surface ponds may be made for the purpose of promoting ground water recharge.

Canals as Water Bodies

Overtime, as population grew and moved away from the rivers, there was imperative need to expand their farmlands as well, and several centuries later to start industries. As there was no river or a water body in these regions, the demand for water was met through the innovation of canals which carried a portion of the water of the river to the required destination. Initially, the canals were just taken out of rivers and they could carry water just as long there was water flowing in the river. These were called 'inundation canals'. Later, the system of building 'anicut' and barrages was introduced in India and the first ever anicut was built on river Kaveri to take its water to irrigate lands far away from the river. The concept found wider acceptance in irrigation engineering and anicuts were built on several rivers in the world.

In more recent times, in the late second half of the 19th century, extensive network of canals was built to take waters of Ganga and Yamuna rivers to far away regions in what now constitute the western Uttar Pradesh, Uttarakhand, Haryana and Punjab. Prominent amongst them are: the Upper Bari Doab Canal (1848) the Upper Ganga Canal (1854), Agra Canal (1874), Lower Ganga Canal (1878), Sirhind Canal (1886) and many others. One of the single largest canal system in the world, the Sharda Canal, (comprising the main canal and its distributaries), covered a total length of 13,624 km was completed in 1928.

In post-Independence India, great priority was accorded to the construction of multi-purpose dams and water reservoirs on the upstream side of the dam and associated with it, a canal system. Besides being effective in controlling floods, irrigation from their waters increased not only the cultivable area but also the farm yields. Additionally, the reservoirs opened new avenues of employment through fish culture and generation of hydroelectricity.

Construction of canals continued unabated in independent India, starting with the Bhakra Canal (1954). Most contemporary among them are the Indira Gandhi Canal and the Narmada Canal. The former carries the water of the Himalayas to dry regions of western Rajasthan and the latter, of central India, through a dam on Narmada river at Navagam, Gujarat, called Sardar Sarovar, to far off dry Saurashtra. Narmada Sagar is one of 30 dams on river Narmada and is among the largest in India. Both these canals have transformed the moribund lands of the two states into fruit baskets for the country and have brought prosperity and heralded new life in far off dry and drought-prone lands of Rajasthan and Gujarat.

The construction of Narmada Sagar Dam was completed under many odds with the so-called environmentalists fighting a pitched battle with the government against its construction. Deciding in favour of the Dam in 2000, the Supreme Court of India observed: 'The argument in favour of the Sardar Sarovar Project is that the benefits are so large that they substantially outweigh the cost of immediate human and environmental disruption. Without the dam, the long term costs for the people would be much greater and lack of an income source for future generations would put increasing pressure on the environment. If the water of Narmada river continues to flow to the sea unused, there appears to be no alternative to escalating human deprivation, particularly in the dry areas of Gujarat'.

'The project has the potential to feed as many as 20 million people, provide domestic and industrial water for about 30 million, employ about one million, and provide valuable peak electric power in an area with high unmet power demand (farm pumps often get only few hours of power per day). It will also provide flood protection to riverine reaches measuring 30,000 ha covering 210 villages and Bharuch city and a population of 4 lakh in Gujarat. In addition, recent research shows substantial economic multiplier effects (investment and employment triggered by development) from irrigation development. Set against the future of about 70,000 project affected people, even without the multiplier effect, the ratio of beneficiaries to affected persons is over 100:1'.

The Narmada canal system introduced paradigm shifts in canal management system. The canal surface was covered with photo-voltaic solar cells to generate electricity reducing at the same time the loss of water through evaporation.

Canal irrigation dominated the Indian scenario until the 1960s but in 1970s, it yielded first place to irrigation with ground water through both dug and bore wells. Lest it sounds that the canal system is hegemony of north India, it may be stated that the hard rocky plateau of the south is not amenable for the building of canals but canals have been dug in the delta region of Kaveri or in the coastal regions.

National Plan to Inter-Link Major River Basins

It is recognized that (i) the normal rainfall in several parts of the country is inadequate to sustain farming and even that it is uncertain and unreliable, (ii) the per capita availability of water in the different river basins across the country is not equitable (iii) there is always

Figure 7

The ambitious inter-linking concept of Indian rivers

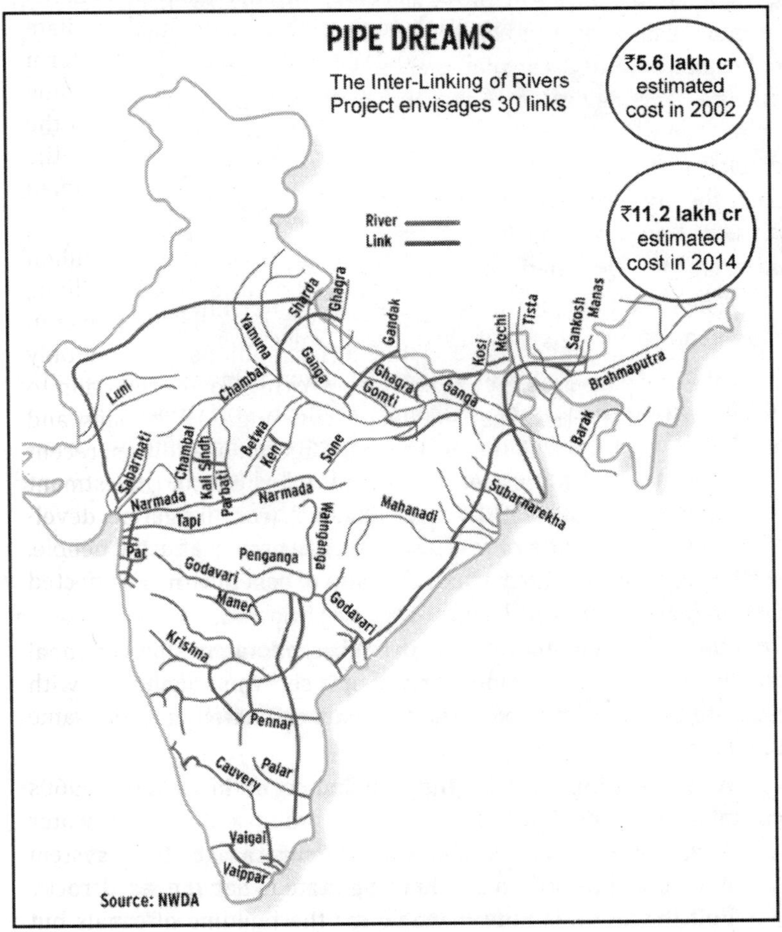

looming uncertainty about both the quantity/quality and timing of the monsoon rains and (iv) while statistical data for the entire country may show average rainfall, there may be local regions acute drought, or floods. Water management experts had been advocating for long time the need to interconnect the river basins, so that water from the 'excess' region may be transferred to the region where it is needed the most. They cite the following as the accruing advantages of the concept: (i) it would facilitate management of situations of both floods or drought, (ii) ensure increased productivity from both agriculture and horticulture, (iii) would bring increased area under cultivation, (iv) increase the production of hydroelectricity, (v) widely increase the availability of clean drinking water for the rural masses, (vi) promote fish culture and open fresh avenues of employment and (vii) open alternate and cheaper method of transportation of goods by making rivers navigable. However, increasingly louder voices are also being raised against the project on grounds of, (i) it causing displacement of large populations from their traditional habitats, (ii) it causing the already shrinking forest cover of the country to diminish further, thus affecting local flora and fauna, and (iii) irrevocably damaging the local environment and ecological system of the river.

The ambitious features of the river inter-linking project are shown in Figure 7 along with estimated project cost.

Almost two decades ago, the Government of India constituted the National Water Development Agency, NWDA, to study the engineering feasibility of such a gigantic project. In order to make the implementation of the lofty project feasible, NWDA divided the country into three zones and considered it more prudent to first inter-connect the rivers within a specific zone. The three zones are: (i) basins of rivers emanating from the Himalayas, (ii) basins of east flowing rivers of the Deccan plateau and (iii) basins of the west flowing rivers emanating from the central Indian water shed. NWDA generated a plan of 32 link canals for the first two zones. This plan is shown in Figure 8 and the estimated cost of its implementation was estimated at more than ₹ 11.0 lakh crores.

As a beginning, work for the inter-linking of the following basins was taken up in the first phase:

1. Godavari–Krishna rivers in Andhra Pradesh have been successfully linked through a 174 km long canal in September 2015, the Pollavaram Canal; the Pattiseema lift irrigation system pumps

Figure 8

The proposed 32 link canals by the National Water Development Agency

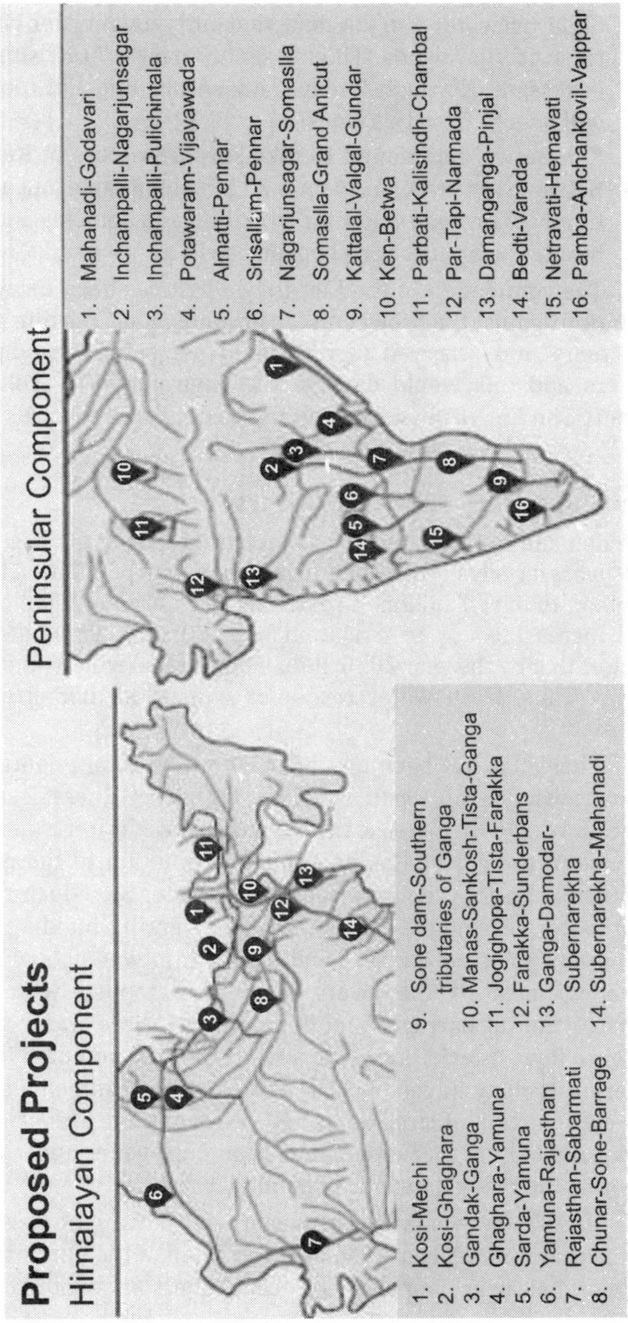

Godavari water into the canal that empties its water into Krishna river at Vijayawada. Thousands of farmers in Krishna, Guntur, Prakasam, Kurnool, Kadapa, Ananatpur and Chittoor districts will benefit from this work.

2. Ken-Betwa Link aimed to take surplus waters of Ken River to Betwa River through a dam, a 231 km long canal and a 2 km long tunnel to provide water for irrigation and domestic applications, besides also generating hydroelectricity.

The completion of the Ken-Betwa link has been delayed due to agitation against the project on the grounds that a portion of the tiger sanctuary and reserved forest would get submerged in the dam waters and this would damage the environmental ecology of the region. The link is now expected to be completed towards the end of 2017.

Vanishing Glory of Pristine Rivers

The demand for fresh water has increased during the past little over 100 years largely because of increase in human population from about 2 billion to over 7 billion, threatening to be around 9 billion by 2050 and increasing use in irrigation and industry. Population experts project that by the year 2030, Indian population would outstrip that of China. Thus, fresh water resources will be further stretched and stressed.

There has not been any increase in the total quantity of fresh water available on earth since its inception. There is no natural process which can increase the amount of fresh water available on the surface of the earth. Mankind sought a solution to the problem by supplementing his requirement of fresh water by reducing the runoff rain water reaching the seas through rivers by building dams and reservoirs to store the seasonal excess rain water. In the past few decades, most of the rivers have been tapped with relentless construction of barrages and dams, leaving little water and silt for onward flow. Interference of this kind holds the potential to affect the riverine ecology adversely. The voices of environmentalists against the construction of large dams and reservoirs are often subdued and subordinated to the anticipated economic gains and the need of meeting the rising demands of potable water.

Until a few years ago, the energetic flowing of clean water in streams and rivers conveyed an impression of the immortality of the flow of their sparkling waters. Recent events have shown that whilst water is immortal, the rivers are not as some of the best known rivers the world over – Colorado in USA, Indus in Pakistan, and Gomati, Yamuna, Krishna and many others in India – are struggling for their survival in order to maintain the grandeur of their riverine ecosystems. The present day highly degenerated state of the water resources could be attributed to:

1. Development of irrigation-based agriculture and horticulture, and excessive use of chemical fertilizers, pesticides and insecticides in order to provide food security to the huge global population.
2. construction of large number of dams and water reservoirs in their flow path which reduce their flow, more often below the critical amount required to maintain their ecosystems.
3. Withdrawal of their waters for industrial applications and discharge of industrial effluents and untreated sewage water.
4. Global warming which has in turn caused the rainfall pattern to become erratic – sometimes too much within a short period, resulting in flash floods and excessive quantity of 'runoff' water and sometimes too little akin to drought conditions. In both cases, the local ecosystem is destroyed.

Traditional Methods of Monsoon Prediction

Traditionally, the Indian farmers have been depending either on the astronomical calculations and deductions provided in the *Panchang* – the Indian almanac – to have fore-knowledge about the timing and strength of the monsoon or on the traditional knowledge conveyed from generation to generation through thumb rules and folk tales. These methods flourished in different forms in all parts of India; perhaps with a fair measure of success. For example, the height at which a particular bird (perhaps the weaver bird) builds its nest in summer in the proximity of a water body, is regarded as providing good indication about the strength of monsoon in many parts of India; if the nest is at a good height, the rains would be copious but if the nest is at low height, the rain is more likely to be scanty. Entomologists believe that insect movement is another sure indicator of monsoon behaviour as birds and insects can sense wind shift, humidity and atmospheric pressure changes through their antenna and these instinctively define their response. Thus, en mass, movement of ant colonies, often carrying their eggs is seen as a precursor of rain.

It is also felt that the flowering or fruiting behaviour of trees could also provide reasonably accurate prediction. Thus, an abundant flowering of neem trees is regarded a sure indicator of good rains. It is said that the Amaltas (Golden Shower) tree reaches flowering peak 45 days earlier than the start of rains.

It is now gradually, but very slowly, being felt that there may be some science behind the traditional methods of forecasting monsoon, as they are based on behavioural patterns of ants, crows, flowers and certain fish species. In a study, sponsored by the Indian Meteorological Department, the non-government organization (NGO), the Vishwa Scientific, Pune tracked the western coast and interviewed a cross section of people in the society to identify the natural indicators commonly used by them to anticipate the behaviour of monsoon. It was found that fishermen in the Konkan area predict the onset of monsoon by the behaviour of small fish, like Etoplus, which move in large numbers to the shore before monsoon sets in. In Melghat, a tribe called Korku believed that a local flower called *Bahava* blooms 40 days prior to the setting in of the monsoon. Similarly, the Bhil tribe in the Thar Desert of Rajasthan regards the flowering of the bushy Khair (*Acacia catechu*) trees and of the wild cucumbers as good omens for good rains.

Literature on traditional knowledge of predicting monsoon is rare to find; some of it is found in the books written by Varahmihir around 700–800 CE or in the rapidly disappearing couplets of Bhadli, composed around 1100 CE, as they usually passed from mouth to mouth. Bhadli – a 10th or 11th century savant – sang many songs about weather. In recent years, academicians at leading agricultural and management institutes have begun to evince great interest in proving the validity, or otherwise, of traditional knowledge in this regard. One of Bhadli couplets states that 'if there is rain accompanied by lightening and roaring of clouds on the second day of Jayastha (Indian month beginning latter half of May), there will be no rain for the next 72 days'. It so happened (according to Professor Kanani) that in the year 1990, the second day of Jayastha fell on 4 June when there was abundant lightning and thunder; and 72 days later was 15 August when the rains came to break the long period of drought. That one data point could well have been a coincident but that year was one of general drought in India. The converse of that couplet, however, has held ground as there has been no rain on the second day of Jayastha since then and no 72-day continuous drought has since been observed. Further, the 8th century book, Brahad Samhita by the philosopher Varahmihar states that monsoon begins 45 days after the peak blooming of the tree Indian Laburnam (*Casia fistula*). Similarly, the blooming time and the quality of the bloom of another plant, Keshuda or 'flame of the forest' (*Butea Monosperma*) are used to predict the possible onset of monsoon. If it blooms early, it indicates early monsoon; late flowering means late monsoon. If it blooms densely, it means the next monsoon will be good. So with this plant the farmers are predicting two different variables – the date and the quality of rain. However, much remains to be done to resuscitate the traditional Indian knowledge and to integrate it with rational scientific principles. But it is not an easy task. Be it as it may, modern science is approaching this important subject of anticipating the monsoons in its own manner.

Modern Scientific Methods of Monsoon Prediction

It was not long ago that it was thought that the strength of Indian monsoon was mainly governed by the thermal conditions prevailing in summer months over the Indian land mass; the hotter the continuous spell of hot weather (uninterrupted by mild showers) during the astrological conditions of the so called *nau tapa* (nine days of intense heat), the better it was for abundant monsoon rains.

It is now generally recognized that the Indian quota of the monsoon is not merely influenced by the thermal conditions prevailing over the landmass and the adjoining seas but several globally inter-connected environmental factors are involved in influencing it. Their study is a matter of deep studies in the field of atmospheric sciences.

The El Nino Current and the Indian Monsoon

The impact of monsoon rains on the national economies of many countries, including India, is so great that there is global effort to understand/identify the environmental and geographical variants that affect its distribution amongst the countries in the path of its movement. Meteorological opinion seems to be converging that amongst the factors influencing its distribution are the impacts of El Nino marine stream of hot water in the Pacific Ocean and of the La Nino effect in the atmosphere above it. El Nino is a stream of hot water, starting off the coast of Peru and moving westwards in the Pacific Ocean and affects the temperature of oceanic surface layers; La Nino effect, on the other hand, affects the atmospheric pressures over the Pacific Ocean. Very broadly, the stronger the El Nino, the weaker is the Indian monsoon but stronger the La Nino, stronger the Indian monsoon. However, the matter is not as simple as this and it is matter of intense studies in a number of countries. The Indian Meteorological Department is developing its own complex model involving a larger of parameters in its efforts to provide reliable guidance to Indian farmers and others.

The trilogy of water resources in India is Monsoon Rains – Surface Waters – Ground Water.

References

(All retrieved, January 2017)

About the Ramsar Convention: www.ramsar.org/about-the-ramsar-convention

Causes, Effects and Types of Sinkholes, Conserve Energy Future, http://www.conserve-energy-future.com/causes-effects-and-types-of-sinkholes.php

China Has Launched the Largest Water-Pipeline Project in History ... http://www.theatlantic.com/international/archive/2014/03/china-has-launched-the-largest-water-pipeline-project-in-history/284300/

China, Tibet, and the Strategic Power of Water, Circle of Blue http://www.circleofblue.org/2008/world/china-tibet-and-the-strategic-power-of-water/

Drainage Systems of India, PMF IAS www.pmfias.com/classification-of-drainage-systems-

Ganges Delta: https://en.wikipedia.org/wiki/Ganges_Delta

Geography of India, Wikipedia https://en.wikipedia.org/wiki/Geography_of_India

Geomorphology and Environmental Sustainability: Felicitation Volume ... https://books.google.co.in/books?isbn=8180690288 – Sharma, & Kalwar – 2005 Concept Publishing Company.

Godavari – India-WRIS www.india-wris.nrsc.gov.in/.../index.php?title=Goda..

Godavari Basin – India-WRIS www.india-wris.nrsc.gov.in/.../index.php?title=Goda..

india-wris.nrsc.gov.in/wrpinfo/?title=Krishna

Kaveri Delta – India Environment Portal | News, reports, documents ... www.indiaenvironmentportal.org.in/.../kaveri-delta/

Know Your Watershed - It's Where Your Water Comes From! — All ... http://allaboutwatersheds.org/library/kyw-poster-files-and-links/know-your-watershed-homepage

Major River Systems of India: India WaterPortal: http://www.india-wris.nrsc.gov.in/wrpinfo/index.php?title=Major_River_System_in_India#

Micro-level Rural Planning: Principles, Methods and Case Studies https://books.google.co.in/books?isbn=8170222966 - R.P. Misra, R.N. Achyutha 1990

Narmada - Tapti - Sabarmati - Mahi - Luni - Ghaggar | PMF IAS www.pmfias.com/west-flowing-rivers-narmada-tapti...

Narmada River - Wikipedia https://en.wikipedia.org/wiki/Narmada_River

Potential Environmental Impacts of Quarrying Stone in Karst - USGS ... https://pubs.usgs.gov/of/2001/ofr-01-0484/ofr-01-0484so.pdf

Publications: Survey of India www.surveyofindia.gov.in/pages/.../59- publications

Ramsar Convention - Wikipedia https://en.wikipedia.org/wiki/Ramsar_Conv

Regional Geography - SABARMATI www.sabarmatiriverfront.com/regional-history

Report of the high level working group on western ghats - Ministry www.moef.gov.in/.../1%20HLWG-Report-Part-1_0...Kasturirangan Committee Report 2013

Report on Water Quality "Hot Spots" in Rivers of India (Central Water Commission, New Delhi, Aug, 2011

River basin atlas of India - A report by Central Water Commission and Indian Space Research Organisation: River basin atlas of India, CWC and ISRO, 2012

Sabarmati Basin - India-WRIS www.india-wris.nrsc.gov.in/.../index.php?... Sabarma...

Sinkholes: The Human Impact: Year In Review 2013 | sinkhole ... https://www.britannica.com/.../Sinkholes-The-Huma...

Suvrat Kher: The Easterly Tilt of Deccan Plateau: http://suvratk.blogspot.in/2011/09/easterly-tilt-of-deccan-plateau.html

Tapti River - Wikipedia https://en.wikipedia.org/wiki/Tapti_River

The Big Melt. - National Geographic Magazine - NGM.com http://ngm.nationalgeographic.com/print/2010/04/tibetan-plateau/larmer-text

The Impact of Human Activities on Dolines (Sinkholes) - Typical ... www.dgt.uns.ac.rs/pannonica/.../volume14_4_1.pdf

The Importance of Watersheds; | The Nature Conservancy http://www.nature.org/ourinitiatives/regions/northamerica/unitedstates/indiana/journeywithnature/watersheds-101.xml

The Ramsar Convention and its mission | Ramsar http://www.ramsar.org/about/the-ramsar-convention-and-its-mission

The Ramsar Convention Manual, 6th edition http://www.ramsar.org/sites/default/files/documents/library/manual6-2013-e.pdf

The Ramsar Sites | Ramsar www.ramsar.org/sites-countries/the-ramsar-sites

The Vindhyan Range: https://en.wikipedia.org/wiki/Vindhya_Range.

Water Clouds on the Tibetan Plateau | Central Tibetan Administration http://tibet.net/2016/05/water-clouds-on-the-tibetan-plateau/

Watershed management in India – Shivaji University www.unishivaji.ac.in/uploads/journal/Journal.../3.pd...

Western Ghats – Wikipedia https://en.wikipedia.org/wiki/Western_Ghats

CHAPTER 7

Properties of Water

Water is the only substance which occurs naturally in all the three states of matter on planet earth, i.e. in solid state as ice, in liquid state as water in lakes and rivers and as vapour (gaseous state) in the atmosphere. All of its three forms had played central role in shaping the topography of the earth over a period of time measured in geological scale. It is both tasteless and odourless liquid elixir of life. In its pure and liquid state, it is clear, transparent and has a light bluish aura around itself. Because of the abundance of water on Earth, its bluish aura is clearly visible when earth is viewed at from space. It earned for earth the distinction of being called the only 'blue planet' in the solar system.

Water is truly formless even as it exists in many forms. In its own massive form, it is seen variously, from tiny dew drops to mighty rivers, to formidable oceans and to awesome ice caps. Apart from its multiple distinctive forms, it is central in shaping the forms of all human and other animate bodies, and of all plants. When present in sugarcane, it is sweet, as sap of a neem tree it is bitter. As amniotic fluid in mother's womb, it protects the growing baby. It helps some amorphous, inorganic compounds of metallic elements to acquire their distinctive crystalline features. Students of science are familiar that the hydrated copper sulphate crystals have a bright blue colour; on heating the crystal, the water of hydration, which had helped the copper sulphate to assume crystalline shape (called the water of crystallization), escapes turning the copper sulphate crystal into an anhydrous white powder.

Water imparts crystalline character to sugar. We are all familiar with crystalline sugar but when its water of crystallization is taken away, it turns amorphous. We are familiar with both forms of sugar.

The latter is called *boora* or *khandasi* and even *desi cheeni* in Hindi. Chemically, sugar is called sucrose consisting of 12 atoms of carbon, 22 atoms of hydrogen and 11 atoms of oxygen; actually sucrose is two simple sugars - fructose and glucose - chemically stuck together. When a saturated solution of crystalline sugar in water is heated up, or cooked, much of the water boils away but the sugar remains in solution and the residual solution is highly viscous (called treacle or *chashini* in Hindi). It is clear that it now has more sugar in solution than what the original super-saturated solution had. When allowed to cool, it may harden into a rather brittle amorphous solid and forms the basis of lollipops, caramels or candy floss.

Water helps seeds to germinate and nurture crops to provide food security to the people. It helps the engineer to drive the wheels of industry. Denial of accessibility to its resources has the potential to trigger animosity between neighbours, internecine violent clashes between communities and even wars between countries sharing a common resource.

It has particularly romantic association with carbon and joins it variously to form organic compounds, including hydrocarbons and various but distinctive petro-fuels. On combusting the petro fuel in internal combustion engines, water is quick to depart from its company with carbon. It comes out as humidity in the hot exhaust gases of jet engines. When the exhaust mixes with the cooler atmosphere in the wake of the aircrafts, it immediately condenses into tiny droplets of water in which form it is visible from the surface of the earth as a white trail, slowly dissipating away in the sky.

Above all, it is water alone that has made the Earth habitable by naturally acting in a number of different ways:

- Acting as a moderator for intense solar radiation and giving earth its characteristic climate and being instrumental in the uninterrupted functioning of the hydrological cycle which ensures continued availability of fresh water through self directed renewals.
- By mechanically cutting or eroding away the mountains creating valleys, and outcrops of porous sedimentary rocks, enabling its own permeation in the strata, maintain the sub-surface soil moist and charting its own flow along with its sediments into streams and rivers which carry them away to distant places creating plains and deltas on their way to the sea.

- Chemically converting the eroded sandy particles of earthy matter into alluvium besides colluding with mechanical forces to facilitate the disintegration of massive rocks.

Water has been able to perform all the above functions as it is endowed by unique physical, chemical and mechanical properties that make it special to the human beings.

What is Water?

Fundamentally, water is a chemical compound which is made when two hydrogen atoms combine with one oxygen atom. This chemical union can occur either explosively or gently. The reaction is highly explosive when an electric spark is passed in a closed space containing a mixture of two volumetric parts of hydrogen with one of oxygen. This very chemical reaction occurs gently when a stream of hydrogen is burnt through a burner in a free atmosphere containing oxygen as in air. A considerable amount of heat is released as a consequence of the chemical union. Efforts are being made all over the world to tame this reaction within engineering possibilities so that hydrogen may be used as environment friendly fuel in internal combustion engines.

With the growth of nuclear thermal power industry, the term 'heavy water' is being increasingly used. Heavy water is water in which the hydrogen atoms are replaced by its heavier isotope, deuterium. It is chemically almost identical to normal water and is used in nuclear industry to slow down electrons. It is not abundantly available in Nature and neither is it the subject matter of our discussion.

Unique Physical Properties of Water

Water has certain properties which have made it a unique creation of Nature. A few, pertinent to the theme of our present submission, are discussed below:

- *Its High Density for a Liquid*: Water has mass or weight (one kg per litre) which makes it move with great force as in a current of a river or as waves in the ocean. We are awed with the destructive power of the Tsunami waves. Such forces in the past millenniums had shaped the land on the Earth's surface.
- *Its High Specific Heat*: When the Sun shines, all substances – metals, organic and inorganic compounds, minerals and rocks – get heated up by the thermal energy of the solar radiation falling on them. But each material requires a different quantity of

thermal energy to raise its temperature by 1°C. This quantity is unique to that material and is called its specific heat. Water has the highest specific heat amongst all materials. As water has a high specific heat, large quantity of heat is required to raise its temperature. One can easily experience the significance of the difference in specific heats of two substances in their response to solar radiation by putting a beaker full of water and equivalent weight of copper or marble in sun for a couple of hours on a hot summer day. Whilst the latter (copper and marble) become unbearably hot, water is just lukewarm. More specifically, if a given quantity of heat raises the temperature of water by 1°C, the temperature of equivalent quantity of copper would rise by about 10°C.

In this manner, the vast oceans store a huge quantity of thermal energy, without making a show of it by way of large temperature rise whereas the landmass does so. This makes the Mediterranean Sea warmer than the Atlantic Ocean because it is surrounded by land mass on all sides and the much hotter North African land mass on its southern boundary.

The evaporation of surface seawater (output) is more than the supply (input) of fresh water by precipitation and coastal drainage systems, causing the salinity of the Mediterranean to be much higher than that of the Atlantic – so much so that the saltier Mediterranean waters sink below the waters incoming from the Atlantic through the Strait of Gibraltar. As water flows into and out of the Mediterranean, two currents are formed in the strait. An upper layer of Atlantic water flows eastward into the sea over a lower layer of saltier and heavier Mediterranean water flowing westward into the ocean, known as the Mediterranean outflow water.

- *Water as Moderator of Earth's Climate*: Innocuous as this difference in response to solar radiation may otherwise appear, it plays and had played important role in shaping the climate and fragmentation of primary rocks on earth. The high specific heat of water makes the vast oceans to act as a sink for solar thermal energy. Acting in this manner, it cools the landmass from getting excessively heated up. In other words, the sea acts as a moderator for the earth's climate. It also means that water releases heat more slowly than others when conditions are reversed.

- *Its Poor Thermal Conductivity*: Whereas metals are generally good conductors of heat, i.e. if one end of a metallic bar is heated, the temperature at the far end also rises quickly; gradually temperature across the whole bar becomes uniform. In contrast, water does not conduct heat as easily. Indeed, water is a poor conductor of heat. Since solar thermal radiation does not penetrate to the depths of water bodies, the temperatures in deep depths are considerably lower than at the surface. Thus, a vertical temperature profile exists is large and deep water bodies. This temperature profile is disturbed in the upper layers of sea water by waves.
- *Its Existence in Three Physical States of Matter*: Water is the only substance that exists in all the three states of matter, i.e. solid (ice), liquid (water as we ordinarily know it) and gaseous (naturally as vapour in atmosphere). Ice exists below 0°C, called the freezing point of water, liquid water in the temperature range 0–100°C and vapour in the form of steam above 100°C which is called the boiling temperature of water. Other substances have also to be heated to change their physical state but to much higher temperatures; thus copper and iron have to be heated to about 1,087°C and 1,536°C respectively to change from solid to liquid state.
- *Water as Transporter of Thermal Energy*: When water is at 0°C, approximately 80 calories of heat have to be taken out from 1 gm of water to convert it to ice at 0°C. Similarly, nearly seven times as much heat has to be supplied to it to convert water at 100°C into steam at 100°C. It is important to note that if water were to change into gaseous (vapour) state at any temperature, thermal energy will have to be supplied to it. When this change occurs through evaporation from the surface of a water body, i.e. without external heating, the vapour would only form by extracting the equivalent amount of heat from the water body, i.e. the body itself would cool down. It is known that rains on land mass come from the condensation of water vapour that had evaporated from the surface of far away seas (largely). The process of condensation would release an equivalent amount of heat wherever vapour condensation occurs in the atmosphere, leading to the formation of clouds. Since evaporation largely occurs on surface of the sea and cloud formation occurs in atmosphere over far off lands, it means that the heat captured by the vapour from the sea is released elsewhere. In other words,

solar energy received in one region on earth is getting redistributed in another, first evaporation and then condensation.

- *Its Expansion on Freezing*: Water exhibits a uniquely different behaviour from other substances when it freezes. In common with other liquids, its density increases as temperature is lowered and is highest at 4°C. Thereafter, the similarity ends. As temperature is lowered further to 0°C, it (the density) decreases but the decrease is very substantial on changing to ice at 0°C; i.e. water expands by about 9 per cent on freezing. Ice, therefore, floats on water. Table 1 gives the density of water in grams per cu. cm at various temperatures in degrees Celsius:

Table 1

Density of water

Temperature (°C)	*Density (g/cc3)*	*Temperature (of supercooled water) (°C)*	*Density (g/cm3)*
30	0.9957	0	0.9998
20	0.9982	–10	0.9982
10	0.9997	–20	0.9935
4	1.0000	–30	0.9839

This means that the density of water chilled at the surface of a fresh water lake increases and it sinks through convection currents and is replaced by the warmer water brought up by the convection currents. The entire water body is, thus, cooled to 4°C. However, when the surface temperature decreases further, it remains on the surface layer which eventually freezes to form ice. Since downward convection of colder water is blocked by the density change, any large body of fresh water frozen in winter will have the coldest water near the surface, away from the lake bed or riverbed. Since ice is also a poor conductor of heat, the frozen surface acts as a blanket to conserve the heat in the unfrozen lake water.

- *The Freezing of Saline Water is Different from Pure Water:* The situation in salt water is somewhat different from the above scenario. It is a matter of common understanding that the density of salt water increases with the salt content in water. It is axiom to state that the temperature at which a solution (e.g. of salt in water, i.e. sea water) begins to freeze is always lower than that of the pure water. Therefore, the total process of freezing of sea water is different from that of freezing of pure water. When a salt solution freezes, three things are simultaneously happening: first, ice of pure water separates from the solution and floats on the surface; second, the salt content in the remaining salt solution increases; and the third, the density of the remaining salt solution increases because of increased concentration of the salt. The increased salt content of oceans lowers the freezing point by about 2°C as well as the temperature of the density maximum of the saline water to be about at the freezing point. Hence, in ocean water, because of the salt content, the downward convection of colder water is *not* blocked by an expansion of water as it becomes colder near the freezing point. Thus, the oceans' cold water near the freezing point continues to sink.

- *Water is a Universal Solvent*: Water is a universal solvent. It is able to dissolve a large number of different chemical compounds – salts, sugar, fertilizers, insecticides, organic liquids and gases. The dissolving power of water is very important for life on Earth. While the solvent property of water has many uses, it has a nuisance value too. Once dissolved in it, water carries all its solutes – chemicals, minerals, nutrients and pesticides, pathogens to wherever it goes, i.e. underground or in surface water rivers/reservoirs. Further, in the course of its slow travel through the grains of sand in the sub-surface naturally occurring elements like arsenic and fluoride also get dissolved in water. Because of this, millions of people in West Bengal, Bangladesh and Nepal are facing the acute problem of arsenic and fluoride toxicity via drinking water from the tube wells.

  Discharge of untreated water from the industry or municipal bodies into water bodies, therefore, causes their chemical and bacterial pollution. Additionally, if the discharged industrial waste water is at higher than ambient temperatures, it introduces another parameter, that of thermal pollution – to the pollution of water bodies. Invariably, this affects the local flora and fauna adversely.

- *Water Has High Surface Tension*: It is an inherent property of water molecules (arising out of its electronic structure) that they are strongly attracted to one another. In other words, the molecules at the surface of the water 'stick together' which act in the manner of 'skin'. We express this scientifically by stating that water has high surface tension. This fact has several implications: (a) The 'skin' is strong enough to support very light objects; some insects (water striders) can walk over it; (b) Water tends to aggregate in drops rather than spread over a surface as a thin sheet; (c) The retention of moisture in soil, against gravity, is the most significant sequel to this property; it is this water that feeds the roots of trees; as one molecule moves up the tree root or through the capillary, it 'pulls' the others with it and (d) The strong intermolecular cohesive forces hold the water column together in the xylem and prevent rupture of the column by the pull due to the transpiration pull. The only other liquid substance, the surface tension of which exceeds that of water is mercury, approximately seven times as much.

The Duality of its Chemical Behaviour

The chemical behaviour of pure water is regarded neutral, i.e. neither alkaline nor acidic. However, in reactions with strong acids, it acts as a base and in reactions with strong base, it acts as an acid. In scientific language, the alkalinity or acidity of an aqueous solution is expressed by stating its pH on a numeric scale on which a neutral solution has a pH of about 7; that of alkaline solutions > 7 and of acidic solutions < 7. On this scale, the pH of lemon juice is about 2.4, vinegar 2.9, apple juice 3.5 and milk 6.5.

Although the tiny droplets of water condensed from the atmospheric moisture, which constitute the clouds, may be initially neutral in character, drops of rain water become acidic as they pick up or dissolve carbon dioxide and sulphur dioxide which are present in the atmosphere.

Solubility of Carbon Dioxide in Water

Carbon dioxide and sulphur dioxide have always been present in the atmosphere but since the beginning of the 20th century, the content of carbon dioxide has been continually rising. Its concentration at the close of 18th century was 0.024 per cent which is now approaching levels of 0.04 per cent. The increase is largely attributed to the combustion of coal and petro-fuels. Its concentration in the atmosphere is now so large that the natural phenomenon of photosynthesis cannot remove it fully from atmosphere to maintain the sustainability of the carbon cycle. Its concentration in atmosphere is, therefore, continually rising. Being a green house gas, its increasing concentration in the atmosphere has given rise to global warming. It is estimated that almost 30–40 per cent of the carbon dioxide, generated globally, enters the sea through runoff waters from the landmass or directly picked up by the sea water through the intermingling of its waves with air. The ultimate destination of atmospheric carbon dioxide is the sea with consequent change in its alkaline character. It is believed that the pH of sea water in the year 1751 was 8.25 which dropped to 8.14 in the year 1994 and the declining trend is continuing. This is called the 'acidification of sea' although the sea water is not yet truly acidic. Even this slight change in the alkalinity of sea water has begun to affect marine ecology, particularly in tropical seas.

The solubility of any gas in water, or sea water, depends both on temperature and pressure over it. The solubility decreases with increasing temperature but increases with increasing pressure. As the temperature of sea waters in the climatically temperate regions are generally lower than those in the equatorial and tropical regions, they would be able to dissolve relatively larger quantities of carbon dioxide. It is a fact of Nature that fathoms deeper into the seas, the temperatures drop and the hydrostatic pressures increase; both the factors promote increasing solubility of the gases. The effects of both the factors, of pressure and temperature, on solubility of gases can be easily demonstrated, and in fact, are matters of common experience. The common soda water, containing dissolved carbon dioxide, is bottled under pressure. No sooner than the bottle is opened (thereby the pressure over the soda water decreases to the level of atmosphere), its bubbles begin to escape in the atmosphere. Similarly when water, saturated with dissolved air or carbon dioxide, is heated in a vessel, gas bubbles will be found forming and rising.

Since sea water is constantly in a state of agitation through the movement of strong inter-ocean currents, convection currents and thermohaline circulation, the pH of any sea is typically its own, unless disturbed by industrial activities, prominently gaseous discharges into the atmosphere and liquids, including hot water, in the coastal areas (thermohaline circulation is a part of large-scale ocean circulation that is driven by global density gradients created by surface heat and fresh water influxes').

Marine Ecosystem as Sink for Carbon Dioxide

Unbelievable as it may sound, the sea hosts a vibrant ecosystem with a variety of life more diverse than available on the landmass. Some of the ecosystems, specific to the coastal regions in tropical and sub-tropical regions, are designed to remove soluble calcium salts and dissolved carbon dioxide through biological activity in the form of calcium carbonate. Many types of marine organisms, without backbones, secrete calcium carbonate which forms a hard protective layer or exoskeleton to protect their bodies. They are commonly found around shorelines of beaches where they are washed ashore by tides, almost always emptied of the organism that had once lived in them.

Mankind has been familiar with the exoskeletons from times immemorial and has been variously using them, including as currency (cowries) or medium of exchange of goods in trade and even as jewellery. In modern times, sea shells are being increasingly used as soil re-conditioners to raise the pH value of the acidic soil, in cement industry and even as a component of poultry feed. The reverberating sound produced by blowing *shankhs* (conch shell) is considered auspicious in many Hindu and Buddhist rituals. So holy is the conch shell in Hinduism that when Lord Krishna blows His Panchjanja *shankh*, its deep sound strikes terror in enemy ranks. Over mining of sea shells from coastal areas for industrial applications is now posing grave threat to marine ecosystems, besides acidification and general warming up of ocean waters.

A specific ecosystem of corals has evolved which consumes the dissolved carbon dioxide. Truly, coral is the most diverse community in marine ecosystem, of the same type as the sea shells. This ecosystem is very sensitive to several extraneous factors, such as the temperature in their habitat, its degree of alkanity or its pH value, clarity and other pollution causing parameters. Ocean scientists have found that the trend towards acidification of the oceans, as a consequence of increased concentration of carbon dioxide in the atmosphere, has adversely affected the coral system. As a result thereof, the physical existence of some coral reefs and atolls is threatened. Another study showed that the very prominent El Nino ocean current in 2010 had raised the sea water temperature enough to destroy almost 16 per cent of coral activity in its path.

What Makes Water a Unique Material?

According to the basic laws of physical sciences, chemical compounds are formed when two elements join together by one element giving electrons from its outer most shell to the other which accommodates them in its own outermost shell (this is the so-called 'ionic bond'; the other kind of bond is the covalent bond). Atoms consist of electrons surrounding a nucleus that contains protons and neutrons. Neutrons are neutral but protons have a relative positive charge of +1 and electrons, a negative charge of –1. Protons and neutrons mainly contribute to the mass or weight of the atom. The electrons are distributed in a number of circular shells around the nucleus but only those which are in the outermost shell take part in the 'give-take' process which gives rise to compound formation.

One measure of the ability of an element to form a compound with another is defined by its electro-negativity. The greater the electro-negativity, the more the atom attracts electrons from the other element towards it or stronger is its urge to form compounds with other elements. On a scale commonly used in chemistry, fluorine is the most electro-negative element, oxygen atom less so. Also, the more protons an element has, greater is its 'pull' on the electrons of the other element, almost like heavier heavenly bodies having greater gravitational pull than smaller bodies. The gravitational pull of moon is one-sixth of the earth's pull. In terms of this concept, oxygen has 6 electrons in its outer most shell with atomic weight of 16 and hydrogen just one electron with atomic weight of one.

As stated above, water is a compound of two hydrogen atoms with one of oxygen. There is a difference in the electro-negativity of oxygen and hydrogen atoms. The highly electro-negative oxygen atom attracts electrons or negative charge (hydrogen) to it, making the region around the oxygen atom more negative than areas around the two hydrogen atoms but the side of the molecule with the oxygen atom has a partial negative charge. They attach to the top of the molecule, as shown in Figure 1. This molecular structure gives the water molecule its lopsided electrical charge that attracts other atoms. A molecule with such a charge difference is called a dipole and shown in Figure 1 and water is called a polar compound. It may be noted that one side of the

Figure 1

Three-dimensional structure of water. The hydrogen atoms are both on the same side of the oxygen atom (black), so the charge is not evenly distributed.

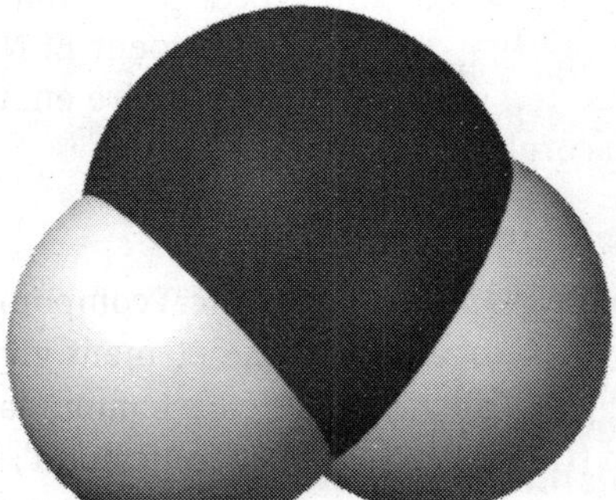

Credit: Ben Mills

molecule with the two hydrogen atoms is positively charged and the other side with oxygen is negatively charged. Since opposite electrical charges attract each other, as do the north and south poles of magnets, molecules of water are connected with one another as shown in Figure 2. This kind of bond is called the hydrogen bond. Though it is much weaker than either the ionic or the covalent bonds, it confers on water unique physical and chemical properties. Some of them have been already discussed above.

Figure 2

Joining together of water molecules through the weak hydrogen bond. The positive side of the oxygen atom in the water molecule joins with the negative side of the hydrogen atom of its next neighbour water molecule.

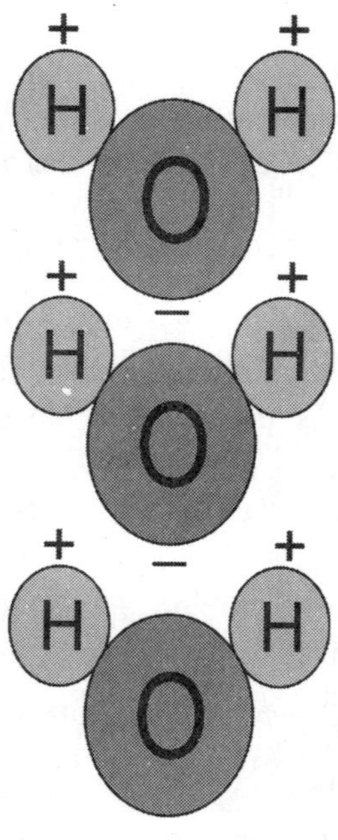

Water is able to act as both acid and base in reactions. For this kind of its duality in chemical behaviour, water is called an 'amphoteric' compound. When reacting with a stronger acid, water acts as a base; it receives an H^+ ion from HCl in the equilibrium:

$HCl + H_2O \rightleftharpoons H_3O^+ + Cl^-$ (Here water is acting as a base, by receiving an H^+ ion.)

When reacting with a stronger base, ammonia NH_3, it acts as an acid. NH_3 by donating an H^+ ion, and is thus acting as an acid:

$NH_3 + H_2O \rightleftharpoons NH_4^+ + OH^-$

(In terms of electronic structure, an acid is defined as a species which *donates* a proton, an H^+ ion, in a reaction, and a base as one which *receives* a proton*)*.

Water as a Chemical in the Formation of Soil from Mechanically and Chemically Disintegrated Rocks

Soils may be regarded as the skin of the surface of earth, somewhere thin and elsewhere thick, particularly in river valleys or as very thick bed when it filled the deeps of the Tethys Sea. Along with water and air, soil is the third vital natural resource without which the Earth would simply have been a rocky planet. Soils provide best habitats for water which are rechargeable or renewable through rains. As we have seen, these habitats occur both below and above the saturated water table.

We have learnt in a previous chapter that at the beginning of the earth, soil was not present. Formation of soil involved first the disintegration of massive rocks through mechanical and chemical weathering processes and carryover of the rock debris or detritus (gravel, sand, silt or other material produced by mechanical erosion and chemical weathering) by flowing water stream to faraway places. During the transportation of the debris, water was not chemically neutral. Blessed by its extraordinary molecular structure, water reacted chemically with the debris to convert it into new types of chemical entities, called soil. Truly, soil is the collective or generic name given to an aggregate of altered minerals in the formation of which water had played a central role. In some cases, the reaction caused hydration. In others the alterations in the chemical nature of the debris/detritus were achieved through processes involving hydrolysis, carbonation or simply dissolving away of the soluble component in the debris. When the organic detritus of plants and living micro organisms that was being carried away by flowing waters got mixed with the soil, the organic matter enriched soil got the distinctive name of alluvial soil but the term 'alluvial deposit' could be used for any sediment (gravel, silt and clay) carried by rushing streams and deposited elsewhere where the stream slows down.

Soils have mostly originated from the weathering of sedimentary which are most abundant on the surface and are mostly made of silicate minerals. In the terminology of mineralogy, the silicate minerals are of various types and called by different names: prominently feldspars (almost 50%), quartz, pyroxenes, amphiboles, micas and clay minerals, The response of different rocks to processes of weathering is different from one another and is closely related to the mineral composition of the rock, e.g. quartz is slow to weather in comparison with pyroxene. The chemical alteration of feldspar in presence of water leads to the formation of a distinctly new clay mineral. There are different types of rocks in different parts of India. Therefore, different types of soil occur in India. Nonetheless, the soil always carried its 'DNA' with it. The soils were classified on this basis in an earlier chapter as alluvial soil, black cotton soil or lateritic soil.

Soil chemistry is in itself a vast subject of study in which the bi-polar nature of the water molecule plays an important role. It makes them adhere to hydrophilic surfaces. Silica and silicates, the major constituents of the Earth's crust, have hydrophilic surfaces and water easily adheres to them. They travel long distances with water currents.

A few of the reactions are involved in the formation of soils. They are indicated below but these may be ignored by those not familiar with chemistry.

Hydration: It is one of the most common processes in nature and involves chemical combination of water molecules with those of the mineral resulting in the formation of a new compound. As a consequence of hydration, there is swelling and increase in volume. It is a reversible process. Wet or moist soils shrink on drying and prominent shrinkage cracks are seen on the surface of black soils under conditions of drought. A few examples of involved chemistry are:

(a) $2Fe_2O_3$ (Hematite –red) + 3HOH → $2Fe_2O_3.3H_2O$ (Limonite –yellow)

(b) Al_2O_3 (Bauxite) + 3HOH → $Al_2O_33H_2O$ (Hydrated Aluminium Oxide)

(c) $CaSO_4$ (Andrite) + 2HO → $CaSO_{4.}2H_2O$ (Gypsum)

(d) $3(Mgo.FeO.SiO_2)$ Olivine + 2HOH → $3MgO.2SiO_{2.}2H_2O$ (Serpentine) + SiO2 + 3H2O

Hydrolysis: Water acts as a weak acid on silicate minerals and is an important process in their chemical weathering. It dissociates into H+ and OH – ions which chemically combine with minerals and bring about changes such as decomposition of crystalline structure and formation of new compounds:

$KAlSi_3O_8$ (Orthoclase) + HOH → $HAlSi_3O_8$ (Silt) + KOH

$HAlSi_3O_8$ + 8HOH → $Al_2O_3.3H_2O$ (Hydrated Aluminium Oxide) + $6H_2SiO_3$ (Silicic Acid)

This reaction is important because the reaction products are available to plants and soluble silica or silicic acid is available for plant growth.

Dissolution: It has been stated that rocks may contain water soluble minerals as well. These minerals get slowly dissolved away in the water running over or through them, leaving their host rock with hollow spaces or rough surfaces. This physical process is intensified in water, made acidic by dissolution of organic and inorganic acids produced during decay of organic matter.

Carbonation: Rain drops pick up carbon dioxide, present in the atmosphere, during their drop from the clouds to the Earth and become acidic through the formation of carbonic acid. Carbonic acid attacks many rocks, specially limestone and brings them into solution. When the carbonate mineral is present as a binder for the sand particles, its removal leads to the disintegration of the rock.

Rainfall occurs over the oceans as well. On account of increasing concentration of carbon dioxide in the global atmosphere, the acidity of rain water is increasing. The ocean waters are trending towards acidification. The change is significant as it has threatened the existence of islands made predominantly of carbonate minerals, such as the coral islands or coral reefs.

Effect of Atmospheric Pressure on Boiling Point of Water

Before the advent of pressure cookers, cooking food at high altitudes in the mountain habitats of mankind took a longer time than at lower altitudes. Science explained this reality by stating that as water begins to boil at temperatures < 100°C at high altitudes, cooking took a longer heating time. In other words, external pressure on water in the cooking pan influences the temperature at which water begins to boil.

Air has weight. Therefore, a column of air on the sea surface would assert greater pressure than at high altitudes simply because of the reduced height of the air column at higher altitudes. The weight of air above a surface exerts pressure on it. It is a physically measureable quantity and can be expressed as force exerted by the air column on a unit area of the body. It is commonly measured by mercury barometer and expressed in terms of the height of the mercury column in a closed one-ended tube, which outside air pressure would support. Explicitly, when a long glass tube, > 780 mm full of mercury is inverted over a pool of mercury, some mercury flows out of the tube; but soon the height of the mercury column stabilizes itself at a certain level. At this level, the pressure exerted by the column of mercury at its surface in its pool is equal to the pressure exerted by the column of air on the surface outside. Under normal atmospheric conditions at sea level, this height is 760 mm. This height is the barometric pressure of the place where it is measured. Obviously, the barometric pressure decreases at higher altitudes.

Water boils at 100°C at sea level but at lower temperatures at great altitudes. For example, on the top of Mt. Everest water boils at about 68°C. We can see this happen. Take a flask, fill it half with water and bring it to boil; keep boiling for sometime so that air is completely driven out of the flask. Stop boiling but put a tight fitting cork to the flask, inverting it, put it on a tripod stand. As it cools, the steam in the vacant space in the flask condenses into water, creating vacuum inside. One would notice that the boiling continues even as the temperature has fallen down considerably below 100°C. Relationship between boiling temperature and pressure in the boiling chamber is given in Table 2. This phenomenon of steam formation at lower temperatures under conditions of vacuum has been widely put to use in desalination of sea water through the thermal route.

Table 2

Relationship between boiling temperature of water and the state of vacuum inside the boiling chamber

Barometric Pressure; mm of Mercury Column	*Boiling Temp °C*	*Barometric Pressure; mm of Mercury Column*	*Boiling Temp °C*
760	100	92.46	50
535	96.11	55.12	40
525.5	90.00	31.75	30
355	80.00	25.40	26.67
233.68	70	22.86	24.44
149.35	60	20.32	22.22

Conversely, at higher pressures, the boiling temperature is raised above 100°C. Thus, water, deep in the ocean near geothermal vents, can reach temperatures of hundreds of degrees and remain liquid. Similarly, in modern high pressure steam boilers, largely used in thermal power generation, steam formation occurs at temperatures considerably higher than 100°C at which temperature water boils under one atmosphere pressure; under pressure of 10 atmospheres, the boiling temperature is raised approximately to 180°C and under 30 atmospheres to about 233°C.

Steam and Water Vapour

In gaseous state water has two forms: water vapour and steam. There are subtle differences between the two. These are: (a) water vapour forms naturally through evaporation at ambient temperatures; the thermal energy required for conversion of liquid state into gaseous one is supplied by the residual water which thereby gets cooled down. This incidentally explains why water in a porous earthen pot cools down whereas water contained in a metallic vessel does not, other conditions remaining identical, (b) thermal energy has to be externally supplied to water for the formation of steam at 100°C; the external source could be combustion of any kind of fuel or solar energy, and (c) whereas steam can exist independently above 100°C in a closed vessel, water vapour cannot exist independently; it co-exists in air but on equal terms with it, i.e. as vapour enters air, an equivalent number of air molecules get out of it. In other words, heavier molecules of nitrogen and oxygen are replaced by the lighter water molecules. Therefore, any moist air is lighter than dry air. Therefore, air after picking up the evaporation from water bodies rises up in the atmosphere. This is the modest beginning for the vital role that water vapour in atmosphere plays in determining the weather and climate of any place and is particularly true for the Indian climate.

The entry of water vapour in the atmosphere is different from that of the entry of a solute, like sugar, in solvent – water. Water vapour enters the atmosphere by displacing an equal number of molecules of air; it retains its independent identity and contributes to the atmospheric pressure. Strictly, the atmospheric pressure exerted at a place by saturated moist air would be less than that of dry air, the other parameters remaining identical. Since atmospheric pressure can be regarded as a measure of the weight of an air column, we may state that the atmospheric pressure at any place is the sum of the weight of the air molecules (nitrogen and oxygen) and the water molecules in the column. Air is, thus, a commonwealth of nitrogen, oxygen and water vapour, each can act independently of the other. Thus, oxygen supports combustion but nitrogen remains indifferent. Similarly, water vapour follows its own rules; it may condense out of air, depending on conditions of pressure and temperature of the atmosphere.

Water Vapour in Atmosphere

Water vapour present in the atmosphere comes from a number of different sources, predominantly from evaporation from the surface of seas and in appreciable quantities from the transpiration from trees and to which a minor contribution is also made by the moisture contained in the exhaled breath or drying of the perspiration. The amount of water vapour present in the air is expressed quantitatively as g/m^3 (grams/cu. m). It is called the absolute humidity. Its value varies from almost zero in the air over the cold Atacama non-polar desert, west of Andes Mountains in South America, to 40 g/m^3 in equatorial and hot tropical regions. Generally speaking, it is high at higher temperatures and low at lower temperatures, if water is otherwise available for evaporation. There is a limit to the amount of water vapour that can remain in air at any temperature (and atmospheric pressure). When a volume of air at a given temperature holds the maximum amount of water vapour, the air is said to be saturated. Air at higher temperatures, can hold greater amount of water vapour; i.e. the net amount of water vapour that air can hold at a temperature of say, 35°C, is considerably higher than what it can hold at any lower temperature. The absolute amount of water that can be held by air at different temperatures is shown in Table 3.

In some respects, the solubility of water vapour in air is similar to the solution of sugar or of common salt in liquid water. There is a maximum limit of sugar or common salt that can go into solution in water and that their solubilities increase with temperature. Similarly, the amount of water vapour that has the potential to enter air increases with the temperature of air.

In short, it varies from place to place and from time to time of the year. It is high near sea but low in far off regions; even near sea, it is less when environmental temperatures are low and more with higher temperatures.

Table 3

Saturated vapour pressure and saturated density of water vapour at different temperatures

Temperature °C	*Saturation Pressure mm of Mercury Column*	*Density kg/cu. m*
0	4.6	0.005
10	9.2	0.009
20	17.4	0.017
30	31.7	0.030
40	55.1	0.051
50	92.2	0.083
70	233	0.20
90	525	0.42
100	758	0.50

The presence of water vapour in atmosphere has made life possible on earth. Its presence in atmosphere affects crops and orchards favourably but in its absence, dry hot winds damage the crops. Dry air makes us feel uncomfortable too by making our skin dry. Above all, water vapour in the atmosphere acts as a regulator or moderator for global climate or surface temperatures by absorbing solar radiation, controls rate of evaporation, releases latent heat on precipitation and plays a central role in defining earth's climate. The presence of water vapour in atmosphere or atmospheric water vapour is an important factor in regulating the climate of a place.

How does Water Vapour Form? The Process of Evaporation

Water vapour forms at the surface of water bodies through the process of evaporation which occurs at all places, at all times and at all temperatures. The conversion of water from the liquid state to gaseous state requires considerable amount of heat, called the latent heat of vaporization. The rate of evaporation depends prominently on (a) proximity to water bodies, (b) ambient temperatures, it is faster in dry and hot summers than in winters, (c) humidity already present in the atmosphere, i.e. aridity or dryness of the atmosphere increases the rate of evaporation, and the rate is slow in humid atmospheres, as in the rainy season, and (d) wind blowing over the water body increases the rate of evaporation; if there is no wind, the air which overlies the water surface gets saturated through evaporation but is not swept away and further evaporation stops until the saturated layer is removed and replaced with dry air.

Water bodies take the required heat or thermal energy for vaporization from the incoming solar radiation. This amount is very large. Almost 600 times as much heat is required to convert 1 gm of water into vapour as is required to raise its temperature by 1°C. Therefore, the higher the amount of moisture in the air, the more is the consumption of solar thermal radiation.

Furthermore, water vapour is the ultimate source of all kinds of precipitation on earth – as dew, fog, rain, snow and hail. When the water vapour condenses into tiny droplets to constitute the clouds, or to get precipitated in other forms, the thermal energy earlier absorbed at the surface of water bodies is released up in the atmosphere. Acting in this manner, the water vapour (–) condensation cycle is redistributing the solar thermal radiation. Water vapour, thus, plays a significant role in shaping the weather and climate on Earth.

The Process of Condensation

Table 3 shows that the amount of vapour required to saturate the air increases with temperature. Viewing the situation conversely, it indicates that the unsaturated air of a higher temperature could become saturated at a lower temperature. The temperature at which this occurs is called the 'dew point'. Or, the temperature at which a given sample of air becomes fully saturated is called the dew point or saturation point. If the temperature is brought down any further, the moisture in excess of the saturation limit would condense out of the air in the form of tiny droplets. Explicitly, condensation of water vapour can occur only below the dew point. The loss in temperature with increasing altitude of the rising moist air occurs naturally. Wherever condensation occurs, it would liberate or release the heat it had absorbed for its vaporization earlier.

Forms of Condensed Moisture

When air happens to hold water vapour in excess of its saturation limit, or above 100 per cent relative humidity, the excess amount tends to precipitate out. A volume of unsaturated air can become saturated prominently in three ways: (a) by cooling the air to a temperature at which the amount of water vapour held at the higher temperature becomes saturated for the lower temperature, (b) by mixing of two masses of unsaturated air of different temperatures when the water vapour content of the mixture at its new homogenized temperature is higher than the saturation level at that temperature, and (c) by increasing the input of water vapour into the volume of air (may be in a limited volume of air).

The first two occur naturally in atmosphere. The atmospheric temperature decreases naturally as the altitude above the sea level increases. According to International Civil Aviation Organization, for every increase of 1,000 m altitude, the temperature drops by 6.49°C. Along with drop in temperature, atmospheric pressure also decreases at increasing altitudes; this is shown in Table 2. As the moist air rises up, it could achieve a height where the temperature is so low that the water vapour content for saturation at this temperature is less than the amount of water vapour the air was holding when it was at the surface level; the air then at once becomes super saturated with respect to water vapour. The same phenomenon occurs naturally when moist air rises up on the wind facing side of the mountains. The resulting super saturation causes the so-called orographic precipitation (also called relief precipitation) of rain in higher valleys. The phenomenon listed in (b) could occur suddenly under cyclonic conditions with the arrival of a cooler mass of air from higher levels of the atmosphere and its mixing with the air with higher content of water vapour at lower altitudes.

The third of the above listed methods of creating super saturation of water vapour in air forms the basis of upcoming innovative methods of recovering fresh water from sea water and is discussed in the chapter on water harvesting.

Water vapour may condense either as droplets of liquid water (dew, mist or tiny droplets constituting the clouds) or as tiny particles of its solid form (snow, frost, hail). The former type of condensation occurs when the dew point is above the freezing temperature of water and the latter, when it is lower.

Condensation as Dew: Water vapour deposits as dew when objects on the earth's surface, such as grass blades, leaves of plants and stones, become cooler than the atmospheric air and their temperature falls below the dew point. The phenomenon of dew formation from water vapour can be physically seen when a bottle cooled in a refrigerator is brought out; the outer cold surface of the bottle brings the temperature of the air in its contact down below the dew point, and moisture from the air condenses on to it.

Condensation as Frost: When the dew point is below freezing point and the temperatures of the surface bodies also falls below the freezing point (but not that of the atmosphere), the condensation of extra moisture could take place in the form of very minute particles of ice crystals, called frost.

Condensation as Mist and Fog: When condensation takes place in the air near the earth's surface in the form of tiny droplets of water hanging and floating in the air, it is called mist.

Condensation to Form Clouds: Higher above, where the atmospheric temperature is less than the dew point, water vapour may condense in the form of very tiny droplets, < 0.02 mm diameter, which remain suspended there and become visible from the ground as 'clouds'.

Relative Humidity

Since the amount of water vapour at any given place varies from time to time, meteorologists or weather scientists use the term 'relative humidity' which defines the water vapour content of the air relative to its content at saturation level. It is the ratio of the amount of water vapour actually in a volume occupied by air to the amount the space could contain at saturation. On this consideration, the relative humidity of an air at saturation point is 100 per cent. If the relative humidity of a place is less than 100 per cent, the air is said to be unsaturated. We are told by metereologists that near the surface of the Earth, the relative humidity rarely falls below 30 per cent. If a given sample of air is holding 10 g of water vapour at a temperature at which the saturation limit is 20 g, its relative humidity is 50 per cent. For a given sample of air, its relative humidity increases, when the temperature of the air goes down, and vice versa too; its relative humidity could also increase with the addition of or mixing with moist, the absolute humidity of which is greater than its own.

The difference between the two terms - absolute and relative humidity - may be made clear by the following statement. The air masses over the tropical deserts, such as the Thar Desert in Rajasthan or the Sahara and Namibia Deserts in Africa, hold vast quantities of moisture; but because of the high temperatures, however, the relative humidities are very low. Conversely, air in very high latitudes, because of low temperatures, is frequently saturated even though the absolute amount of moisture in the air is low. In tropical deserts, there is a large difference between the temperatures of day and night. Whenever the night temperature falls below the dew point, dew may settle down to provide solace to whatever kind of surviving organic life there. Strange as it may otherwise sound, Nature has created life to survive under such extreme climatic conditions. A good example of this kind of life is beetle, named Stenocara. This beetle is native to Namibia Desert in southern Africa which is one of the most arid and hottest regions of the world, receiving annually only 14 mm of rain. The beetle is able to survive by collecting water on its bumpy back from early morning fogs, formed by the condensation of the humidity, triggered by the drop in night temperatures in sandy deserts. To capture this water the beetle stands on its legs on a small ridge of sand, facing the breeze at 45E and catches the tiny droplets of water present in the fog on its wings. As morning fog sweeps across the desert, water droplets stick on the peaks of the bumps, eventually coalescing into a bigger drop which rolls down straight into its mouth. Scientists are trying to mimic its shell's natural architecture to design efficient techniques of harvesting water from dry atmospheres or to improve visibility at airports by clearing fog by collecting the water present in the aerosol of fog.

Condensation of Atmospheric Moisture and its Impact on Climate

As stated earlier, the moist air is lighter than dry air; it rises up to the cooler zones in the atmosphere. As it does so, the gravitational force gradually weakens, as a consequence of which air expands. The process of expansion consumes energy which the humid air is obliged to provide and in the process of doing so, itself cools down. It, thus, augments the cooling due to increase in altitude of the uprising air. The converse of the phenomenological happening can be demonstrated by compressing air in a bicycle pump which gets heated up as the pumping process is continued. As the moist air rises, at some altitude, it gets sufficiently cooled down to cause nucleation of micro-sized droplets of water. The process releases the heat of vaporization. Every gram of water condensed releases about 600 calories of heat – and another 80 calories of heat per gram of water could be added if the temperature up there is low enough to freeze the condensed drops in the upper atmosphere. Enormous amount of thermal energy, thus injected into the cloud imparts to it dynamism. By warming the humid air, it may help the cloud to grow and move upward. It may even re-evaporate the droplets if the temperature of the warmed up air is above the saturation temperature.

It is apparent that the higher the vapour content of the atmosphere, the more latent heat available on its condensation to promote dynamism or turbulence in the atmosphere. When the updraft continues due to increased temperature, the cloud mass may continue to grow in size, even to the extent of forming a thunderstorm cloud, and may rise to great heights, may be up to about 12,000 m, propelled by the thermal energy and humidity variations at altitudes. The upward movement of the, thus, formed thunderstorm cloud simultaneously induces a downward movement of the upper cooler air. When the updrafts are particularly strong, a vortex may develop in the turbulent air, just like a suction vortex forms in the pool of water in a wash basin when water is allowed to escape through a pipe at its bottom or when water poured in a funnel gushes out through the funnel bottom. Vortex formation in the earth's atmosphere is an important phenomenon in meteorology.

An analogous situation created in the atmosphere when the upward movement of a thundercloud, activated by a huge release of thermal energy, creates a downward swirling movement of air with

Figure 3

Photograph of a cyclone over Iceland on 4 September 2003

Source: International Space Research Station.

wind speeds in the range of 300 to 600 km/h. If the vortex touches ground, the speed of the whirling wind (as well as the updraft and the pressure differences) can cause tremendous damage, tearing apart homes and flinging potentially lethal debris. A tornado or a cyclone is, therefore, a narrow, violently rotating column of air that extends from the base of a thunderstorm to the ground. Since the cyclone formed by the air is laced with tiny particles of ice (coming as it does from high altitudes), it becomes visible to the camera (Figure 3). The vortex of the cyclone is clearly seen in Figure 3.

Cyclones can occur almost anywhere in the world but more frequently over USA, Argentina, Bangladesh and Bay of Bengal and affect the coastal states of India.

References

Chemical Weathering of Rocks http://www.agriinfo.in/?page=topic&superid=4&topicid=249:

Gentle Introduction to Water: http://www.chem1.com/acad/sci/aboutwater. htm

How Soils Form; Queensland State Government: https://www.qld.gov.au/environment/land/soil/soil-explained/forms/

K.R. Arora, Formation of Soils, http://www.civileblog.com/formation-of-soil/

Physical Properties of Water: http://www.physicalgeography.net/fundamentals/8a.html

Water Molecule: University of Idaho: http://www.webpages.uidaho.edu/geol309/new_page_4.htm

CHAPTER 8

Water Harvesting

Apart from making quantum jumps in several scientific and technological achievements, including geophysical and space exploration, nuclear and solar energies to cite just a few, the latter half of the 20th century will go down in history during which there was (a) explosive growth in consumption of water due to, per se, massive increase in population, and its expanding use in industry and irrigation, (b) deepening water table of ground water resources all over the world and near exhaustion of few fossil aquifers in desert countries, (c) pollution of surface and ground water resources, (d) rising economic activity, population boom and climate change, (e) neglect and damage to natural habitats of water through human activities and (f) awakening of social consciousness amongst enlightened people the world over to provide securities of food, sanitation and health, gainful employment to every individual of the global population, crossing over 7 billion, which resulted in creating a heavy demand for fresh water as its easy availability was central to the success of lofty ideals. Scientists at the famed Massachusetts Institute of Technology, USA, are making dreary forecasts that countries in Asia, including India, may face severe water shortages by the year 2050.

Progressively increasing, the demand for water has now become so high that it cannot be met from the annually renewable surface water resources. The shortfall was initially met through exploitation of ground water resources through shallow dug wells all over the world. As the demand increased, technology helped mankind to pull out water electro-mechanically, which Nature had hid deep down within its bowels. As a result thereof, even the arid regions of the world began to bloom with human population and sandy desert lands turned into green crop lands. As the demand for water continued to soar up, these Nature bequeathed underground resources, once thought to be inexhaustible, began to get stressed and some soon got exhausted, with others getting depleted beyond redemption. But about 40 years later, at the beginning of the second decade of the present century, the continued sustainability of the man-made green croplands in deserts entered the grey zone of uncertainty as there were no means to recharge them. Within a short span of about 4–5 decades of the unchecked widespread 'loot' of ground water extraction, the coffers of the earth began running dry. All this brought the earth to the brink of ecological disaster of unprecedented dimensions.

The looming threat of the disastrous consequences of the exhaustion of ground water resources alerted many other countries who were using underground water resources, the state of health of which was defined by the depth of the water table below the habitable surface. Repercussions of depleting water table also began to be felt in many rainfall deficient regions of India, of which the water shortage in Alwar district of Rajasthan is a typical example, which was heightened by the droughts in the 1980s. As it happened, the crisis attracted committed social workers, led by the now globally renowned Rajendra Singh, who single handedly started the movement to give to earth, its share of rain water. The movement was akin in many respects to the earlier 'Chipko Movement' of Sundar Lal Bahuguna for the protection of Himalayan trees wherein people stood around a tree to prevent it from becoming a victim of the axe, in truly Gandhian style.

Soil Demands its Share of Rain Water

The rainfall deficient states of India have had rich traditions of rain water harvesting since the Vedic times. Historically, the archeologists have dated some of these systems to 1500 BC. In the mad rush to lustily grab even the underground resources of fresh water for its ever increasing use for gainful commercial exploitation, mankind began to deny the Earth any share in the bounty of Nature. Disgusted with selfish attitude of mankind, the Earth soon began to express its resentment by causing desertification of surface areas. Thus sending clear warnings that its share of rainwater has to be honourably given to it. Traditional Indian practices had recognized this vital need. Their water harvesting techniques or systems varied across India in accordance or harmony with the local microclimatic and geomorphologic features. We will illustrate this with few typical examples.

The Disappearing Surface and Ground Water Scenario in Rajasthan

Rajasthan is the most water deficient state of India; its western or the Thar part receives less than 400 mm of rainfall annually through the monsoon during June to September, leaving the soil to parch the rest of the year but the eastern part is better endowed with water resources, though yet dependent on ground water resources. In the last few decades, the desert region has seen many changes, including a manifold increase in both human and cattle population, resulting in overgrazing which has accentuated the drought conditions, despite the availability of Himalayan waters through the Indira Gandhi Canal.

Rajasthan has long been known for its underground water resources and droughts. Alwar district may be taken as typical of eastern Rajasthan. Its traditional water harvesting system was through *johads*. A *johad* is essentially a crescent shaped dam of local earth and rocks built at a place with sharp gradient to intercept the runoff rainfall. The concept of *johad* is shown in Figure 1. On the face of it, a *johad* stored water on the surface; its most important function, however, was to recharge the ground water resources by enabling the accumulated water to percolate downwards through

Figure 1

Conceptualization of a johad

From Peter Bosshard blog.

the soil. It is claimed that a *johad* is able to recharge aquifers within a radius of a kilometre. In the pre-1940 period, there was enough ground water for the farmers to grow irrigated crops of wheat, mustard and beans, besides the monsoon irrigated crops. It is documented that at the beginning of the 20th century, 60 per cent of the land was covered with forests. It began to grapple with severe water scarcity and drought conditions in the 1980s. In retrospect, it appears that human interference with Nature triggered the onset of the calamitous situation.

Alwar, then a princely state, allowed massive deforestation in 1940s which turned the revered trees into railway sleepers and charcoal. Whilst this act brought financial rewards to the princely state, it affected the local ecology in several ways: (i) destroyed wild-life, (ii) degraded land and exposed the top soil to erosion; (iii) compaction of the soil under the impact of falling rain drops reduced percolation, (iv) the soil washed down the slopes of barren hills got deposited in the *johad* ponds which over a period of time got fully silted up, and (v) the silt at the bottom of the pond also slowed down the infiltration of water into the aquifer. In course of time, the *johads* began to be neglected. The neglect led to their eventual disappearance. As they vanished, the natural system of ground water recharging also got vanished. Fortunately, the *johads* remained in the faded memories of village elders.

With the dwindling of the surface water resources, people began to meet their requirements through ground water through bore wells, often installed with government support. The decreasing input of fresh water along with increasingly greater output caused the underground water to retreat deeper below the surface. As a consequence, another vicious cycle set in: trees and other plants that depended on underground water died because the water was beyond the reach of their roots. Without the plants protecting the soil, erosion increased and *johad* ponds got filled up more rapidly with silt. In addition, less vegetation meant less transpiration which reduced the humidity in the atmosphere; with dry atmosphere, the summers became hotter and trees withered away. With the drying up of wells and village forests, the source of firewood for domestic hearths also vanished. Women and children had to spend up to 10 hours a day walking to distant locations to fetch firewood and water, hauling the water home in earthen jugs balanced on their heads.

With global warming, the monsoon seasons became shorter, from 101 days in 1973 to 55 days in 1987. The result was even less rainwater to replenish the aquifers. The local seasonal streams and rivers had already begun to dry up, only to remain in the faded memory of village elders. Their memory of the local ecological scenario got triggered when the hitherto unknown, Rajendra Singh got into action to revive the *johads*.

Return of Underground Water: Rajendra Singh and Tarun Bharat Sangh (TBS) Restore Traditional *Johads* for Rain Water Harvesting

Singh and his two colleagues from Tarun Bharat Sangh took up pickaxes and spades and began to dig out one of the *johad* ponds, identified by village elders in the village Gopalpura. Progressively others joined and after seven months, the *johad* had been excavated to a depth of 15 feet. Singh and his colleagues awaited the monsoons. As luck would have it, the monsoon of 1986 was the region's first significant precipitation in four years. By the end of the season, the pond was full. And amazingly a neighbourhood well, that had long been dry, had profuse quantity of water. The event ushered an era of new beginnings.

The success inspired the Gopalpura residents to take up restoration of a crumbling irrigation dam; they reconstructed the 6 m high and 350 m long dam by 1996, and built nine *johads*. As a result, the underground water table rose from about 15 m to 7 m after the rains. Availability of water in the neighbourhood wells saved the women from the labourious trips to distant wells. Moist subsoil allowed crops to thrive with less irrigation. Because well levels were higher, less diesel was needed to pump out water to the surface. The area of wheat fields jumped from 33 to 108 hectares, and some farmers diversified into sugarcane, potatoes and onions as well. Many of their fields could now produce two crops, one in the rainy season and an irrigated crop in the dry season. Encouraged, the residents decided to reforest 10 hectares along the edge of the village to help protect the *johads*, by cutting down on soil erosion.

The movement spread to neighbouring Bhaonta where water was available in only two or three of its 20 wells during the 1985–88 continuous drought years. As a sequel to the restoration of a large number of *johads* in the area, not a single well dried up during the next drought from 1999 to 2002. Significantly, the *johads* revived the dead

Arvari river back to life with the river having water all the year round. After that, other rivers that had gone dry also resumed their year-round flows. But a bigger battle was being fought on a different front.

The residents of the Sariska Tiger Reserve were also building *johads* in the early 1990s but their wells were not filling up. The problem turned out to be illegal marble mines, which were sucking the underground water dry. TBS moved the court to close the mines. More than once, Singh and other organizers were beaten by hired thugs. But the group finally won a Supreme Court order, shutting down 471 marble mining pits. Villagers got their underground water back.

Traditional Sustainable Practices of Rain Water Harvesting in Thar Desert

The Thar Desert of Rajasthan presents a totally different scenario. It is the most populated desert in the world where every drop of rain has to be harvested. Its driest part is in the west but semi-desert conditions prevail in the east. The geomorphology of the western part of Thar Desert in India is characteristically different from its eastern part; it experiences extreme climate, rainfall of even less than about 400 mm annually and strong pre-monsoon winds prevailing over thick land surface of loose sand. The wind lifted sand often accumulates in scattered heaps, called sand dunes; though not every location has sand has. Sand is not the only requirement for the formation of a dune. The formation of sand dunes requires fulfilment of the following conditions:

1. Availability of sand.
2. High wind velocity to lift and carry sand.
3. Some obstacle to break wind velocity, so that a part of the lifted sand particles are deposited around the obstacle, and take the form of a dune; while the rest is carried away with the wind, only to be deposited at the next obstruction. The obstructions are provided by the sparse vegetation of sturdy plants with deep penetrating hardy roots. Such dunes usually have parabolic contours and are anchored at the tips by the obstruction which helps to block or break a part of the wind and the rest moves away, only to be blocked at the next obstruction. A depression is, thus formed between adjacent dunes. It acts just as a river basin does for its river. In this case it collects rain water for downward percolation into the ground. Without plants and their anchoring roots, loose desert soil would be moved easily by near-constant

blowing winds. There would also be no depressed sand bed to collect the rain water, howsoever temporarily.

4. Presence of moisture in the atmosphere. Condensation of moisture in the nights causes sand grains to stick together; larger grains are more difficult to transport as higher wind velocity would be required to lift them.

It is clear that atmospheric humidity helps to stabilize the dunes and depressions, and facilitates plant growth, the roots stabilize the loose sandy sediment and the plant itself breaks the wind velocity. The infiltrating water is collected within unconsolidated/semi-unconsolidated formations or in the faults of consolidated stratum. The ground water occurrences are usually very deep (30–100 m). Taking it out manually with the help of long ropes on pulleys is very labourious and tiresome affair. Thus, even beneath desert sands, there is water, though how deep is its occurrence, is a matter of the local geology of the desert land.

Ordinarily, a sand dune could be mobile because strong winds which prevail before the monsoons could lift sand particles and carry them away. Strong winds have greater power to lift the sand. The amount of sand moved by wind is a power function (exponential factor) of the wind speed. For example, a 10-km-per-hour wind carries 13 grams per hour, a 20-km-per-hour wind carries 274 grams per hour, and a 30-km-per-hour wind carries 1,179 grams per hour (*Encyclopedia Britannica*).

Traditionally, the scanty rain water is harvested by building *taankas* in the depressions at a suitable place along the gradient. A *taanka* is a paved cylindrical or rectangular, underground rain water storage cistern, lined with lime mortar, usually 3–4 m in diameter and as much deep wherein rain water from natural basins (e.g. depressions between sand dunes) or from roof tops and courtyards is collected. Once fully filled up, the water would suffice for a few months for a family of 4–5 persons.

This technique was largely abandoned in the post-1950s as water began to be made available through pipe lines or was locally pumped up. As a consequence of increased human and cattle population, increased withdrawals resulted in depletion of ground water resources and traditional systems began to be revived at the beginning of the present century.

There is heavy loss of water due to evaporation from the exposed water surfaces of the water in harsh desert climate. Being underground and covered, there is minimal evaporative loss from *taanks*. A *nadi* or dug out open pond or *kund* is the oldest and still prevalent method of rain water harvesting.

Stabilization of Sand Dunes

Dunes being naturally mobile, it is clear that the stability of the dune is central to the local availability of water. Efforts are, therefore, directed to stabilize them. All such methods aim to reduce the velocity of the wind. The following three approaches are adopted for this purpose:

- Planting vegetation cover. Studies carried out by Indian scientists show that in low rainfall (150–400 mm/year) regions of the desert, the sub-soil moisture content could be about 5 per cent at a depth between 1.5–4 m during the pre-monsoon months, suggesting that plants could survive. Plantation is least expensive and most commonly used method of stabilizing of sand dunes, besides being aesthetically pleasing. Afforestation with trees, the roots of which reach great depths is preferred so that they can take water from deeper strata.
- Raising of micro wind breaks. Twigs and brush wood are inserted into the ground, parallel to one another about 5 m apart. This practice has the additional advantages of reducing the evaporative loss of soil moisture and increasing its humus content.
- Make-straw checker boards. This is the widely practised Chinese practice wherein wheat or rice straw or reeds are placed vertically in the ground so as to form about 50 cm tall rectangular grid of 1–2 m size.

Glorious Example of Rain Water Harvesting in Thar Desert

A glorious example of rain water harvesting on a large scale in the arid Thar Desert is the Ghadisar Lake, at Jaisalmer. Spread over an area of about 5 km x 1.5 km, it was built during the period 1336–67 CE. Its catchment area is 400 sq. km. Water starts coming into it with the first rain; when it is full, the excess water spills over its weirs into nine other tanks built in the directions of natural slope; any further surplus is directed into wells called *taanka* in local parlance.

Rain Water Harvesting Systems of Chandel/Bundela Kings in Drought Prone Central India

Bundelkhand region spreading over states of Uttar Pradesh and Madhya Pradesh is another nefarious, drought prone area of India.

Since 2007, the region has been facing drought due to failure of successive monsoons. Geographically, it lies between the Ganga alluvial plain to the north and the Vindhya Range to the south. The average annual rainfall in the region is 800–900 mm but during the 2007–14 years, it received only 400–450 mm annually. Besides, the pattern of rainfall has changed, perhaps due to global warming. Normally, the number of rainy days in Bundelkhand is 52 days (Meteorological Department of India) but during the last six years, it was restricted to just 24 days.

Though the region has a number of streams (tributaries of Yamuna river), continuous failure of the monsoons for a few years in a row, has severely restricted water availability in both the surface water (rivers, lakes and ponds) and the ground water systems. The deepening of ground water table led to consequential withering away of forests, and other natural vegetation like fodder grasses. The gently sloping uplands, which were once endowed with rich forests, got bereft of them; the hills are now barren with sparse vegetation. The region now appears to be perennially short of water. The moot question is: is the failure of monsoon alone responsible for this situation or is/are there any other factor/s?

Vagaries of nature in the form of failure of monsoons are not new and history vouches for their occurrences in the past as well. In the historical past, the concept of devising water harvesting systems specifically suiting the geological and topographical features of the micro basins was extensively utilized by Bundela and Chandela rulers of central Bundelkhand region during their reign in the 851–1,545 period. They realized that the streams could not be used to irrigate the farms because their channels were generally below the cultivable land. They overcame the situation by building a very large number of ponds at places where advantage of the natural topography was available. These artificial lakes/ponds were formed by making embankments across the lower extremities of valleys, and thus arresting and impounding the waters flowing through them. At other places, where the rivulet flowed between two hills, they blocked the path of the rivulet by building earthen/lime mortar dam between the hills. It is interesting to note that wherever such dams were built, the sub-surface stratum is found to be a predominantly impermeable quartz rock with unconsolidated structures. Whether the selection of the site for the ponds was based on some understanding of local geology or it was a sheer coincidence that the chosen site was geologically compatible would always remain a mystery.

Many of the ponds and lakes built by the Bundela kings gradually disappeared in the post-1950 years due to illegal encroachment by unsocial elements as the public remained indifferent and silent spectator to their nefarious activities. With reduced surface area consequentially available for infiltration of rain water, the watertable started going deeper down, and caused the hills to become bereft of greenery. The assaults on ponds continued unabated widely and in course of time began to encroach upon the Kishore Sagar Dam in Chattarpur as well in the first decade of the present century. In earlier times, the dam used to meet the demands of water when the other ponds in the vicinity had dried up. According to Central Ground Water Board, Chhatarpur has 143 ponds/lakes which irrigate their neighbouring areas. It is heartening to note that consequent to a petition filed by an enlightened individual (Brij Lal Misra), the MP High Court ordered in September 2013, to get the encroachments on the Kishore Sagar Dam vacated.

The loss of surface water bodies is not a phenomenon confined to any particular region of India. The malaise is widespread all over India. According to data recently (March 2016) released by Centre for Science and Environment, New Delhi, Raipur (Chhattisgarh) fares the worst with 80 per cent of its water bodies destroyed by urbanization, with Bangaluru being the next with 79 per cent, followed by Ghaziabad (UP) and Kolhapur each with 75 per cent, and Guwahati 60 per cent. Whilst it may be too late to recover the lost water bodies, every effort must be made to restore as many of them and to safeguard others through conscious, knowledge-based societal awakening. There is no doubt that public, bereft of knowledge, remained blissfully apathetic even as their lifeline in the form of surface water bodies were getting destroyed or getting choked up with silt and urban waste and willful assaults by land mafia. Lakes also prevent sudden flooding caused by heavy rains as they act as 'sponge' and absorb rain water and prevent development of localised flood like situations.

Rain Water Harvesting Through Temple Tanks and *Baories/Vavs*

Other systems of water harvesting in other parts of India included the traditional water storage tanks in temple campuses or ponds in the vicinity of human habitation. Almost all temples in south India have tanks that are in good shape even today. These stored rain water; some large and deep ones had steps built so that pilgrims could always reach the water.

The realization that the ground water level waxes and wanes seasonally, a unique system was devised so that large number of people could reach the surface of water together. In this a large mass of earth was removed so as to form a very deep stepped tank. These were called *baoris* or *vavs* and were mostly built between the 5th and the 19th century; some are architecturally outstanding with exquisite stone carvings. Besides providing drinking water, *baoris* were meeting place for festivals and other celebrations.

Tanks have been the lifeline of the semi-arid Telangana state since the 11th century when the Kakatiya rulers, built a very large number of tanks, estimated to be in excess of 14,500 and thus recognised the role of local topography in the storing of monsoon rain water in tanks. Tanks such as Ramappa, Pakhala, Laknavaram, Ghanapuram and Bayyaram helped agriculture and overall development and prosperity of the region. Many of the tanks got destroyed due to societal neglect. The government is now planning their restoration through desilting and ensuring their protection with the involvement of local people.

Rain Water Harvesting in Other Regions

The urgency of rain water harvesting differs for the regions according to the climate and rainfall pattern. For example, in central Kerala, small heaps of local soils were made before the end of winter monsoon; water collected in the deeps surrounding the heaps got time to infiltrate effectively into the ground, with the result that there was no serious shortage during the scorching summer months before the onset of the monsoon in early June.

Similarly, in areca nut-growing part of Karnataka, husk was spread evenly on the ground to retain water, both by absorbing it and retaining it within its loose packing. Water, thus, got time to infiltrate into the soil to become part of the ground water. In a way, this technique mimics the process adopted by Nature in the dense forests in the Western Ghats. It is reported that under pressures of increasing labour costs, such environment friendly methods are becoming less popular. The fact that underground water could be pulled out through deeper bore wells seems to have contributed to the decline of traditional environment friendly practices.

Though there was no special need to harvest rain water in the alluvial plains and river valleys in earlier times, there is urgent need now to do so because of heavy withdrawals of ground water to meet both the agricultural and domestic needs. In recent years, the beverage industry has worsened the ground water scenario at many places because of point source withdrawal of large quantities of water.

Man-Made Lakes/Reservoirs in Historical India

Several lakes were made in accordance with the local topography to collect the surface runoff water. Among the large artificial lakes built in India to harvest rain water, mention may be made of Gujarat's 'Sudarshan' Lake in Girnar built by Mauryan kings, as early as 300 BCE; or the Bhoj Tal built by Raja Bhoj in the middle of 11th century CE at Bhopal. It was built across two hills and fed by 365 streams and springs, including underground streams (aquifers) according to local myths.

The unique system of collecting rain water in manmade lakes in the Mewar region of Rajasthan, nurtured by the Maharanas, (Kings of Mewar were addressed as Maharanas) occupies a prominent place in the annals of glorious Indian traditions of rain water harvesting. As the population of their kingdom grew, so did the number of lakes. In modern times, the selection of sites for large water reservoirs involves geological considerations, one is awe-struck that the then selected lake sites are in complete accord with geological issues involved in selection of sites for perennial lakes. Since modern geological sciences are generally believed to have made a beginning in the 19th century, the methodology then adopted for site selection remains shrouded in mystery.

The topography of the Mewar region is like a huge dish/basin, surrounded by hills – Aravalli Hills (average height 573 m) along the north-west, and on the other sides by the highlands of Gujarat, Malwa of Madhya Pradesh and Rajasthan. Thus, the rain water from a very large part of the country flows into the basin through several streams rivulets and underground channels.

The saga of water harvesting began with the building of the Chittorgarh Fort about the 6th or the 7th century. It is one of the largest forts in India and a World Heritage Site. It is situated on a hill 180 m in height, above the plains of a valley drained by the Berach river. It is said that within the precincts of the fort, there were once 84 water bodies, but only 22 are now surviving. Favoured by the undulating topography of the region, they were/are in the form of ponds, wells and step wells. These water bodies are fed by rainfall in the natural catchment area. They have a combined total storage capacity of 4 billion litres that could meet the water needs of an army of 50,000 for four years.

When the Maharana was forced to abandon the Fort in the 16th century, he chose Udaipur as his new capital. In retrospect, it appears that the selection was again influenced by the availability of water in the surrounding regions. It is amazing that the foundation for building up a series of lakes in and around the present-day city of Udaipur was laid by the vision of an unnamed local youth who had built a small dam on the local rivulet, Sisirma - a tributary of the river Kotra - at a village called Pichauli to provide water for drinking and minor irrigation in 1362. This dam was later strengthened by the Maharana. Almost two centuries later, in the year 1560, Maharana Udai Singh further reinforced it and raised its height. Thus was born the famous Pichaula Lake.

During the period 1680–87, the then Maharana built another earthen *bund*/dam to the north of the Pichaula dam in the Aravalli hills to catch the waters running down the valley slopes of the hills. Being weak, it was washed away by flood waters. It was rebuilt with lime and mortar in 1889. With the new and enlarged dam, the new water body formed was named after him as Fateh Sagar. After the Pichaula Lake, it is second of the four lakes of Udaipur city, the other two being Udaisagar and Dhebar or the Jaisamand Lake.

Keeping the tradition of his predecessors alive, Maharana Raj Singh built the Raj Samand in 1660 on river Gomati; the dam also receives water from two smaller rivers, Kelwa and Tali. On the southern end of the lake, the embankment is built of white marble. It is rather ironical that the marble which earned for it the distinction of being classified as a World Heritage site became the cause of its undoing a little over three centuries later. Indiscriminate mining for marble in its catchment areas caused irreparable damage to the source of its water. As a result, the lake had almost dried up in the year 2000, but efforts are now on to revive it. The lake Pichaula was also trending towards a similar fate.

Alluding the shrinkage of the lakes to excessive evaporation, a layer of cetyl alcohol was spread on the lake surface with a view to prevent the surface of water making interface with the air in the hope of reducing the loss of water due to evaporation. Whilst the results of the experiment were inconclusive, it was clear that the malady lay elsewhere. Similar experiments were tried out elsewhere too in the world, perhaps with no success.

Why Indians drifted away from their glorious traditions of rain water harvesting could be a matter for social scientists to investigate and debate; for us it is more important to revive them and to protect the sanctity of their perennial resources so that future India is as resurgent as it was in its historical past.

The concept of storing water in artificial lakes in the historical past was uniquely different from the present times. The present day practices build dams on the river paths, but the earlier ones did not; most of the lakes they created collected water which was the runoff not otherwise supplementing the water flow in rivers. It was, thus, ecologically friendly and sustainable.

How is the Natural Process of Infiltration Choked

One avowed task before the water management gurus is to ensure adequate recharge so that the output of underground water is more than balanced by input of unpolluted water. Human activities have caused both the choking of the process of infiltration and pollution of the infiltrating water. Prominent anthropogenic activities responsible for creating this imbalance are:

- Reduction in the surface area of the ground from where infiltration could occur through encroachment on their catchment areas, indiscriminate urbanization on lands in the immediate catchment areas of rivers, ponds and destruction of ponds to promote building activities. Developed areas with hard cemented surfaces like roofs, factory sheds, aircraft landing surfaces, etc. either shield the ground from water or are impervious and do not allow water to seep through them into the ground.
- Reduction in the water holding capacity of sub-surface soils on account of densification of soils caused by indiscriminate deforestation and willful destruction of tall grasses, such as *kans.* Fine and deeply penetrating network of their root systems entangles within itself small aggregates of soil particulate matter and the root-soil composite acts as natural sponge. The sponge soaks and retains within itself the percolating water. Thus enabling the process of water percolation to continue for a longer time duration.
- Excessive grazing by cattle on grasslands exposes the bare surface to the erosive action of running water which washes

away the silt to the water bodies. As the silt settles down on the bed of the natural water body, like a pond, or manmade reservoirs, it obstructs the process of infiltration because the permeability of silt is low. Consequently, not only the aquifers are not charged but the water holding capacity of the reservoir is reduced due to continued deposition of the silt.

A good amount of rainfall is, thus, getting wasted as it ultimately goes back to the sea via the rivers by first flowing into the local lower areas. This water is qualified by the term 'runoff' water. The high velocity of its flow in the same channels, consequent to its increased quantity, additionally erodes the banks of streams and *nallas* (gutters on land) and muddies their waters.

Pollution of Runoff and Ground Water

Two basic characteristics of water are responsible for the pollution of its underground resources. One is, its being a universal solvent and second is its natural tendency to infiltrate into the ground. Most pollution is due to biological, industrial and agricultural activities of human habitats. Some of them are listed below:

- As the water travels across open or agricultural fields, roadways, gullies and *nallas*, etc. it picks up remnants of fertilizers, pesticides and insecticides and soluble part of human and pet waste, etc. It infiltrates into the ground with them only to pollute the otherwise pristine ground water.
- Unrestrained discharge of untreated water from sewers, septic tanks into rivers, lakes and wastelands results in their pathogens polluting the ground waters. Already, ground waters in many parts of the world, including China, border of USA/Mexico and India are heavily polluted and unfit for human consumption, without pre-treatment.
- Toxic lagoons created by industrial wastes such as the red mud tanks of primary aluminium industry, or the ash ponds of coal fired thermal power stations or the manure lagoons at pig farms, etc.
- Toxics from solid waste dumps adjoining urban population centres.
- Waste water can also carry as sediments or suspensions of loose soil, lubricant discharges on roads and wash waters from automobiles etc. into ground water.

In short, the immaculate purity of rain water is thereby lost. Pollution of underground water of ground water sources is a serious problem, particularly with developing economies, India included, due to rapid and unplanned urbanization.

Pollution of surface water resources was first noticed in the United States, United Kingdom and countries of Western Europe over 50 years ago. They adopted stringent measures to regulate all kinds of discharges into them as a result of which water bodies have regained their healthy status. River Thames, the pollution of which had assumed as notorious dimensions as that of our rivers, Ganga included, is today clean and abundant with fish. This shows that where society makes water quality a priority issue, many water resources can be made usable once again.

Much of the polluted runoff reaches the water bodies and they too get polluted. The gradual accumulation of insecticides and other toxins in water bodies has now reached such dimensions as to pose serious threat to local flora and fauna, including the humans. The unsavoury fact is everything put or discarded on the ground ends up in water bodies. Even as the use of insecticides and fumigants has improved human health by reducing cases of malaria and other vector transmitted diseases, diseases like gastro enteritis, jaundice, typhoid are on the rise in the population using polluted water.

Minor human activities like watering plants and gardens, car washing, or malfunctioning septic tanks, etc. are equally responsible for water pollution. Therefore, each individual member of the society has a responsibility to make sure that contaminants stay out of our water resources. The move started by the Government of India in 2015 to stop open air defecation is a welcome step in the direction of preventing ground water pollution in India. Large discharges from the industry, however, are easier to control but need strong governmental and legislative support besides public awareness.

Some Consequences of Pollution of Water Bodies

Non-point source stormwater runoff pollution generally occurs anywhere people use or alter the land. Most people are, however, unaware of how they impact water quality. For example, the ingress of fertilizers and nitrates in the lakes through polluted runoff can cause too much algae and water hyacinths to grow, which use up the dissolved oxygen needed by the water body fauna to survive, as a consequence of which the fish die. Similarly, oils, soaps and detergents hurt fish gills and fish skin and other aquatic animals, and other chemicals damage plants and animals when they enter the water. Many insecticides are metabolized by the animals that ingest them but others are not. Some, like mercury compounds discharged either knowingly or unknowingly into water bodies, get accumulated in the body of fish. When the mercury infected fish is consumed by humans, they suffer from neurological problems caused by severe mercury poisoning. So serious has been the problem that the use of mercury has been banned globally, including in medical thermometers. In India, similar problem may arise when people ingest uncooked vegetables, such as tomatoes, onions, carrots, radish, etc. cultivated by irrigating the crop with untreated sewage water as the toxins and pathogens are first assimilated by the crop and then passed on to humans who consume uncooked vegetables.

Even in the scientifically most aware country – the USA – water bodies have been unconsciously affected by pollution. A study (Weekly Reveal) reported that excessive use of pesticides and fumigants in Californian fruit gardens has created pollution of its water bodies and the water is tainted with lead. It is a matter of deep concern that the quality of ground water, now being tapped out from shallow depths from few places in the country is no longer of pristine quality. As the quality of rain water has always been the purest and pristine, it is clear that anthropogenic activities have affected the quality of water infiltrating down and adulterating the ground water. There is a common adage in English language that 'as you sow, so you reap'. This holds literally true for ground water.

Public apathy and indifference towards issues of social relevance is primarily responsible for the pollution of runoff water and this has put India way down in the list of countries in terms of the quality of potable water. The results of a countrywide survey conducted by the Central Pollution Control Board in 2005 showed that ground waters in almost all parts of India were variously chemically polluted. Their data found 66 per cent had unacceptable organic values and 44 per cent samples had coliform from faeces. Widespread contamination of ground water with sulphate, nitrate, salinity fluoride and/or arsenic contamination is a serious health issue in many parts of India. According to the data released by United Nations, globally a child dies under conditions of poor sanitation every 15 seconds on account of non-availability of clean water.

Simple Steps to Prevent Pollution of Water

It is easier to prevent pollution entering the water bodies. It is much more affordable than cleaning the water afterwards. Some suggested steps are:

- Avoid overuse of fertilizers and do not apply them before a heavy rainfall. Use herbal pesticides. If chemical pesticides have to be used, follow the instructions provided by the manufacturer.
- Dispose of household hazardous waste properly. Electrical equipment often contains chemicals such as lead and mercury which, if not disposed of carefully, can get into water bodies and harm people, aquatic life and animals. Fluorescent tubes/bulbs contain mercury; in the absence of any organized collection system, the fused ones are often picked up by rag pickers, broken for their metallic caps, thus littering the ground with their fluorescent coating mercury, the ultimate destination of which are the water bodies through runoff water. Lead and phosphorus pentachloride are other toxic compounds released as a result of breaking of cathode ray tubes.
- Recycle all used motor oil. Do not dump used motor oil in drains or on the ground as it contains toxic chemicals that are harmful to humans, animals and aquatic life alike.
- Do not throw untreated sewage water in rivers and other water bodies.

In actuality, the itemized list of steps to avoid pollution of water bodies is endless. There is no alternative to every individual's personal and conscious efforts in this direction.

What is, and Why Rain Water Harvesting?

We have seen that rainwater harvesting was being practiced in India for a very long time and was the lifeline of the desert and drought prone regions of the country. Water harvesting is the process of augmenting the natural infiltration of rainwater into the underground formations by some artificial method/s. A number of factors contribute to the amount of water harvested. These are: quantity of rainfall, the quantum of runoff water and the ease with which the rain water percolates through the subsoil to recharge the ground water. Rain water infiltration into the ground (or harvesting) occurs naturally in open rural areas. But in congested, over-paved metropolitan cities, we need to create methods to capture the rainwater and to charge it into the ground.

The following are some of the important reasons for promoting rain water harvesting:

- To reduce the surface runoff of fresh rain water to the sea.
- To augment the natural processes of ground water recharge and thus arrest the water table from receding further down.
- To arrest the decline of ground water quality by arresting ingress of saline water or water containing harmful fluorides and/or salts of arsenic.
- To reduce soil erosion.
- To inculcate the culture of water conservation and to regard water as a valuable resource.
- To inculcate among the masses respect and reverence for water, so that the sanctity of its resources is intuitively maintained and their destruction or pollution avoided.

Advantages of Rain Water Harvesting

There are several gains from rain water harvesting. A few are listed below:

- It is a way to capture the rain water where it falls and either store it on surface or recharge it into the ground, before it gets polluted.
- It is a simple, eco-friendly method of water conservation and is a necessity in deserts and many other areas.
- Reduces demand on ground water. The stored water could be easily used without any treatment, for several purposes, e.g. flushing toilets, washing clothes, watering gardens, etc. thus reducing load on the municipal supplies of clean water.
- Reduces soil erosion by reducing the amount of surface runoff water flow, that otherwise would have contributed to flash floods, carry the eroded silt from the surface to the bed of the local rivulet, thereby reducing its permeability.
- The stored water is substantially free of chemicals and pathogenic bacteria that otherwise would have percolated down with rain water to pollute the underground aquifers.
- The overall cost of their installation and operation is much lower than that of water purifying or pumping systems. The systems can be installed with locally available materials and technical expertise.

How to Harvest Rain Water

There are basically two methods of rain water harvesting:

- Roof Top Rain Water Harvesting (RTRWH).
- Surface Runoff Harvesting to recharge the ground water.

Though not strictly rain water harvesting, we may add recycling of used water, including sewage water treatment, after appropriate treatment, as this effectively results in water conservation. In order to ensure quality of recycled water, high level of removal of organic and specific inorganic pollutants from the waste water has to be ensured through processes involving biological and physico-chemical treatment of waste water and finally the disinfection of treated wastewater. This water can be safely used for several urban purposes such as landscape irrigation, toilet flushing, fire protection, construction industry, parks, ponds, fountains and wet lands, etc.

In the organized sector, the Indian Railways are one of the large users of water, particularly for washing and cleaning of trains, etc. The Railways have now begun to use treated recycled water for such purposes.

Typical Roof Top Rain Water Harvesting (RTRWH) System

In Roof Top Rain Water Harvesting (RTRWH), the roof becomes the catchment area of rain water. The catchment surface area for rain water harvesting could be any flat surface provided by a flat/sloping roof, terrace, paved floor or open ground. The rainfall on the defined surface is directed to the collecting system through an arrangement of down take water pipes or drains. The harvested water is either stored in a tank, with or without percolation system to purify it or diverted to a recharge system like pits, trenches, dug wells, hand pumps and recharge wells and shafts. A typical roof top water harvesting system is schematically shown in Figure 2. The method is not expensive and if implemented by all the households in a residential area, it can be very effective in augmenting the ground water level in the zone.

It is advisable to put a wire mesh at the mouth of each drain so as to restrict the entry of any floating or sedimentary material into the system. Usually provision is made in the pipe system to flush off the water received through the first shower as it may be polluted with bird droppings on the roof or other contaminants which may have collected on the roof during the long dry months. It could be easily anticipated that higher the annual rainfall and greater the catchment area, greater is the volume of water harvested.

Figure 2

Schematic arrangement in roof top water harvesting system

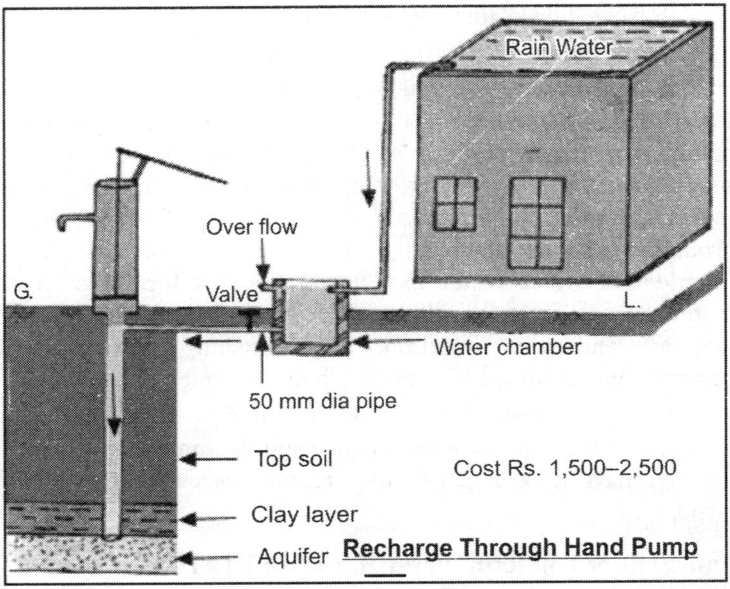

Source: http://www.waterharvesters.com/rain-water-harvesting.

The rain water may be first sent to a filter bed or percolation system, constructed either in a drum or built by brick masonry and back filled by layers of pebbles, gravel, sand and charcoal (which absorbs odour, if any). Each layer may be separated from the other by wire mesh. The primary function of the filter is to remove turbidity, and micro-organisms.

We have used two different terms for almost similar process – infiltration and percolation. However, there is a slight difference between the two. Percolation is a process in which water is made to pass through a bed consisting of different types of filters. On the other hand, infiltration refers to the process in which rain water is absorbed by the soil and continues moving deeper down.

Conceptual Models of Surface Runoff Water Recharging Systems

In situations where the main object of rain water harvesting is to augment the ground water resources, there are three possibilities for ground water recharge as shown in Figure 3. These are:

- Charge the collected water into a basin to recharge a shallow aquifer. The basin could be an existing dug well or pond or could be a specifically made trench or pit for this purpose. These could be back filled boulders, gravels and coarse sand to filter the rain water prior to its recharge into the ground. Hand pumps could also be used as shown in the figure.
- Charge into the ground directly through a deep shaft or recharge well. A recharge shaft may be 100–300 mm diameter; they are generally used to recharge the deep aquifers. Water is often prior passed through filter media to avoid choking of pipes.
- Charge into the deep seated aquifers which are located below clayey surface through injection wells or recharge shafts. Sometimes the same shaft is used to recharge the deeper as well as the upper shallow aquifer through lateral shafts spreading horizontally into the aquifer stratum.

Figure 3

Showing three possibilities of large scale recharging of ground water: (a) a recharge basin, (b) through a well dug into the vadose zone and (c) a direct injection well, deeper than the aquitard

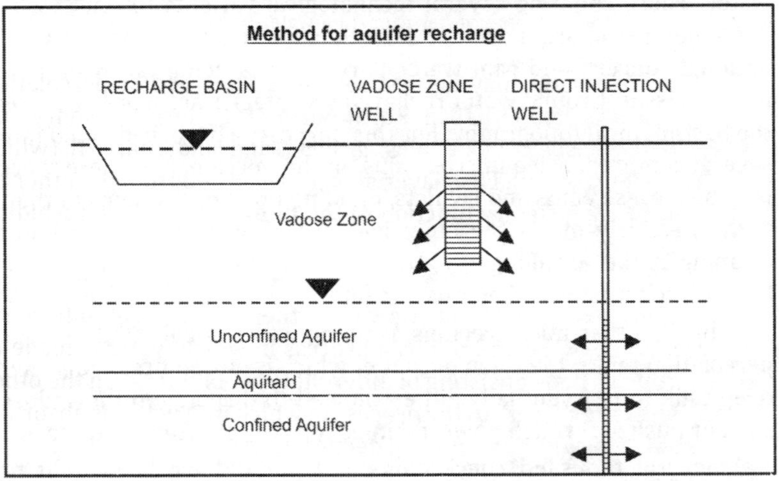

Important question now is that how deep should the shaft be dug into the ground. Conceptually, it should be sufficiently deep to enter the vadose zone, so that rain water exiting from its holes spreads into the ground, without interruption. On an empirical basis, it should extend more than 3 m into the ground to avoid a situation of water logging from developing. In this regard, certain basic facts may be recapitulated.

Water infiltrates into the ground through the pore spaces in the ground strata. In unconsolidated stratum, such as sand, the infiltration is fast and much slower in silted stratum. Truthfully, the depth of penetration should be decided in consultation with hydro-geologists. The geological formations for situations (a) and (b) of Figure 3 may be mutually contradictory in a situation when the pond is required *only* for the purpose of storing water for use; but not otherwise; (this, i.e. the latter situation shown in Figure 3). Therefore, if a pond is required at a place where a layer of sand exists immediately underneath the top soil, the scheme would not work as water rapidly infiltrates downwards. But this situation around the shaft in the vadose zone is welcome. Where the surface runoff rain water is designed to be stored in an open pond, the pond bed should be made impervious with a layer of clay at the top of its bed or through some other means.

The Concept of Micro Basin for Ground Water Recharging

We have presented the basic considerations in the harvesting of rain water. There could be several technological variants for successful implementation. Ground water depletion has now become an issue of national concern and rain water harvesting is being taken up as a national issue. Ground water recharging strategies are worked out on the basis of local topography. For this purpose, a basin is divided into several comprehensive micro basins. In the micro basin where there may be local streams and rivulets, building up of check dams in their path effectively meets the objectives of the mission. This includes cleaning up the bed along its entire course, removing silt and accumulated plastic wastes and encroachments.

In the other micro regions, running water is collected in large percolation tanks, the clean output of which is stored first in another deep tank to be eventually either allowed to disperse in the vadose zone or pushed through powerful injection pumps into the aquifers.

Several thousand check dams and ponds have been built or revived in different parts of India, Karnataka, Maharashtra, Saurashtra and Rajasthan and others. Tamil Nadu was the first state in India to have made roof top harvesting compulsory since about a decade back and improvements both in the water table and quality of water are noticeable.

Emerging Trends: Harvesting Water from Humid Air

There are some dry or mainly dry arid (arid or semi arid) regions in the world where there is practically no rainfall. Yet they are populated. The supply of drinking water to their populace is an existential problem. The ground water is often of poor quality or salty or if potable, it is found at great depths; even so, its level is either continuously falling or the source has emptied out. Lima, capital of Bolivia, is situated at the northern edge of the Atacama desert facing the Pacific Ocean. With practically no rainfall, the air over it is still humid because of its closeness to sea. The atmosphere, therefore, contains substantial quantities of water in vapour form; humidity in summer days could be as high as 90 per cent. Similarly, in Negev Desert of Israel, annual mean humidity of 64 per cent at 19.5°C suggests that the air holds water, about 11.5 ml/m^3 of air.

The threatened global warming is suggestive of temperatures of air rising over the oceans. This would encourage faster evaporation and increased moisture content in the atmosphere. Meteorological data shows that the El Nino warming up of the Pacific Ocean in the year 2015 was one of the strongest. Weather experts attribute the exceptionally heavy rainfall in 2015 in Chile to the increase in temperature of air over the Pacific Ocean.

Scientists in many parts of the world are engaged in developing technology to harvest the water vapour of the humid atmosphere into potable water because there is almost always a small amount of water in the desert air that can be extracted. This is particularly relevant for those desert regions which are close to the sea. These efforts follow either of the following two routes:

- Cooling the ambient air below its dew point and collect the condensed water (the cooling condensation process).
- Absorb the atmospheric humidity in a hygroscopic chemical and recover water through technological means.

New Paradigm in Water Harvesting: The Kalpsar Project

India has launched the most ambitious project for harvesting and storing rain water within the sea. This project is called the Kalpasar Project and involves building a giant 64 km long dam across the Gulf of Khambat from Ghogha in Bhavnagar district to Hansot in Bharuch district. The Dam will effectively isolate the Gulf from the Arabian Sea and would be receiving water inflows of the rivers. Narmada, Dhadhar, Mahi, Sabarmati and some of the small rivers of Saurashtra region. The annual rainfall in the catchment areas of these rivers varies between 400 mm to 1,100 mm. Figure 4 shows the project layout. In fact, the mouth of the river Narmada is to the south of the Dam but its waters would be brought into the Dam through a diversion canal and building a barrage on the river at a place called Bhadbhut. It is hoped that in course of few years, the saline water in the sheltered part of the Gulf of Khambat would turn non-saline due to continuous input of fresh water through the rivers and the canal.

Figure 4

The Kalpsar project layout: the word Kalpsar originates from the mythological Kalpavriksha – a tree that fulfils all wishes.

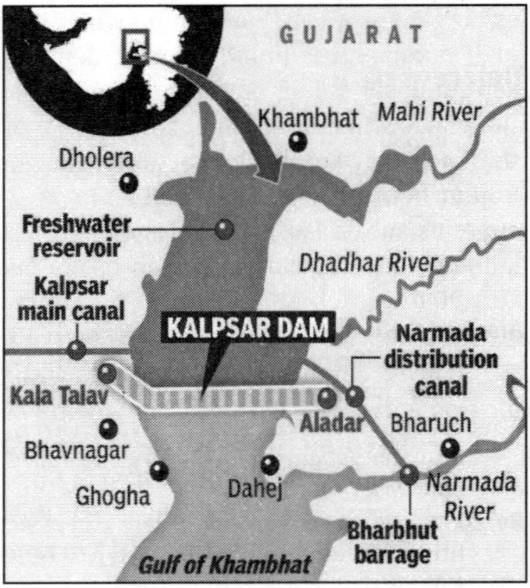

Creation of the fresh water reservoir does not involve any kind of land acquisition or issues connected with rehabilitation of displaced persons or apprehensions of seismic activity. When completed, it would be the world's only reservoir of fresh water in the sea of mammoth dimensions.

The project lists several ecofriendly gains some of which are:

- Saline ground water of coastal areas of Saurashtra and central Gujarat will get converted into fresh water with reduction in soil fertility and about 2 lakh hectares of land along the periphery of the reservoir will be opened for development towards value-based land utilization.
- Fresh water from the huge reservoir will be used to provide water for drinking, irrigation and industrial uses, including fisheries.
- A 10-lane road link will also be set up over the dam, greatly reducing the distance between Saurashtra and south Gujarat.

- There would be a large saving in travel time and fuel due to reduction of distance between Bhavnagar and Surat.

The Kalpsar Project is the most ambitious dream project of resurgent India.

References

2025 Global Warming, Climate Change Predictions 2025 http://www.global-warming-forecasts.com/2025-climate-change-global-warming-2025.php

Akash Vashistha: http://www.dailymail.co.uk/indiahome/indianews/article-2774163/Groundwater-reserves-hit-critical-level- extraction-rockets.html

Amanda Suutari & Gerry Marten: Water Warriors –Rain Water Harvesting http://ecotippingpoints.org/our-stories/indepth/india-rajasthan-rain water-harvest-restoration-groundwater- S johad.html

Atmospheric Water Generator: Water from Air http://www.calvin.edu/academic/engineering/2011-12-team5/downloads/Team_5_Design_Report.pdf

By 2050 Asia at high risk of severe water shortages: MIT study | Reuters http://www.reuters.com/article/us-asia-water-crisis-tracked-idUSKCN0XB207

Chitra Krishnan: Irrigation Infrastructure http://www.indiawaterportal.org/sites/indiawaterportal.org/files/Irrigation-Infrastructure.pdf

Fateh Sagar Lake – Wikipedia https://en.wikipedia.org/wiki/Fateh_Sagar_Lake

GULF OF KHAMBHAT DEVELOPMENT PROJECT (WR). Kalpasar as a prospective Water solution. Saurashtra. A Truly Matchless Project. MISSION. VISION. [PDF]KALPASAR

Kalpasar Project – Wikipedia: https://en.wikipedia.org/.../Kalpasar_Pr...

Lake Pichola – Wikipedia https://en.wikipedia.org/wiki/Lake_Pichola

Large Dams in India: http://www.india-wris.nrsc.gov.in/wrpinfo/index.php?title=Large_Dams_in_India

Large Reservoirs by Volume https://en.wikipedia.org/wiki/List_of_reservoirs_by volume

Maggy B – M. I. T Energy and Climate Control Outlook-2014 http://globalchange.mit.edu/newsevents/news/news_id/413#.WByEsNJ97IU)

Manual for Artificial Recharge of Rain Water: Central Ground Water Board http://cgwb.gov.in/documents/Manual%20on%20Artificial%20Recharge%20of%20Ground%20Water Atmospheric Water Generator: http://www.wikiwand.com/en/Atmospheric_water_generator

Meenakshi Sharma and Vinayak Damle: http://www.sabmiller.in/pdf/ground_water.pdf

Peter Bosshard blog @PeterBosshard

Presentation in english, – KALPASAR: https://kalpasar.gujarat.gov.in/pdf.../Kal...

Rajendra Singh http://www.cgwb.gov.in/documents/papers/incidpapers/Paper%205%20- %20Rajendra%20Singh.pdf

Roopal Suhag: Overview of Ground Water in India; February 2016

Saudi Arabia – Food and Agriculture Organization of the United Nations, www.fao.org/docrep/.../v9978e0k.htm

Saudi Arabia ends domestic wheat production, World Grain www.world-grain.com/...Grain.../Saudi...

Saudi Arabia scraps wheat growing to save water, Reuters www.reuters.com/.../idUSL08699206

Saudi Arabia: Grain and Feed Annual, USDA Foreign Agricultural ... http://www.fas.usda.gov/data/saudi-arabia-grain-and-feed-annual-0

Saudi Wells Running Dry — of Water — Spell End of Desert Wheat ... http://www.fas.usda.gov/data/saudi-arabia-grain-and-feed-annual-0

Sustainable Sanitation and Water Management: Rain water Harvesting (Urban): http://www.sswm.info/category/implementation-tools/water-sources/hardware/precipitation-harvesting/rainwater-harvesting-u

Traditional Water Harvesting Systems of – CPREECwww.cpreec.org/pubbook-traditional.htm

Udaipur – Wikipedia https://en.wikipedia.org/wiki/Udaipur

Water scarcity – Wikipediahttps://en.wikipedia.org/wiki/Water_scarcity

Chapter 9

Desalination of Sea Water for Potable Use

In many parts of the world – including those that have surface fresh water resources in the form of rivers or lakes and potable underground water and those that have arid or semi arid climates – fresh water has become a scarce commodity due to population growth, commodification of water as an ingredient for industrial growth and its unsustainable use in irrigation and dairy farming. Such countries are spread between populated deserts (e.g. countries of the Arab peninsula) and densely populated countries (viz., China, India, Pakistan, Bangladesh, Indonesia and USA). They are said to be water stressed. The term water crisis labels a situation where the available resources of potable, unpolluted water are less than the region's demand. Experts classify water scarcity into two groups: (i) that arising from physical unavailability of water from natural water sources and that (ii) created by mismanagement of available water resources due to extensive use of water for agro-economic activities. Scientists and technologists the world over are variously busy in supplementing the resources of fresh water.

Centuries ago, there was no shortage of water in the populated areas of the world, but water scarcity was felt by mariners of yore. There is historical evidence to suggest that sailors were using simple boilers for raising the steam from sea water, and condensing it to form fresh water. French explorer Jean De Lery reported the successful distillation of seawater during a voyage to Brazil in 1565, and James Cook (1768–1771) desalinated sea water during his circumnavigation of the world. In modern times, however, this concept, based on physical sciences has been integrated with principles of biological sciences to produce potable water from sea and sewage waters commercially on large scale. The success of the engineering has been made possible by materials sciences which mimicked nature to produce permeable membranes. Today's scientific advances in water desalination technology have dramatically increased our ability to transform sea water into fresh water and quench the thirst of 1.2 billion people facing shortages of water. Modern desalination plants are multi dimensional and are no longer based solely on distillation technology. Sea water desalination has become the lifeline of many people and there is no alternative to seawater desalination.

In case of emergency, the mariners also knew that dew condensing on objects from the morning mist comes from the water vapour contained in the atmosphere. They were, however, unaware about the methods to enhance both the rates of evaporation and condensation. The natural process of evaporation from surface of water bodies is a very slow process, the concept remained dormant until methods to accentuate the process of evaporation through inputs of thermal energy were available.

Distillation is an energy intensive process; hence also cost intensive. Since the beginning of its use in small ocean going ships, technological developments in process technology had to await for several centuries until thermal energy was either available for free as a waste from other industries or was available at economically competitive prices towards the middle of the 20th century. Although the technology for the economic production of potable water through desalination of sea water was developed in the western industrialized countries, the first beneficiaries were the desert bound, water deficient, but petro-rich countries – Saudi Arab, Kuwait, United Arab Emirates, Oman, Libya and others. They provided financial filip for the scaling of the technology to commercial scales. In order to convert their natural endowment of petroleum resources into wealth, these countries were obliged to import technology, plant and machinery from the industrialised countries and on increasing numbers of ex-patriot labour. Since their own natural resources of fresh water were not even adequate to support their own native population, they had no alternative but to produce potable water through cost intensive thermal processes.

It is a coincidence that around the same time, the USA, technologically most advanced and resource rich country, with the highest per capita consumption of water, also began to feel the pangs of water scarcity in its arid or semi-arid state of California, because its own resources of ground water had begun to dry up on account of relentless withdrawal of water from them.

Modern Processes of Desalination of Sea Water

Modern processes for desalinating sea water fall into three categories:

- Mimicing the natural hydrological cycle by entrapping solar energy in the so-called 'green houses'.
- Obtaining pure water through distillation using thermal energy.
- Forcing the saline water through a semi-permeable membrane that separates water from salt.

In yester years, it was taught in school science that once a substance dissolves in water, it can only be separated by causing the solvent water to evaporate away completely. Today, however, it is not so as almost pure solvent can be separated from a solution with the reverse osmosis.

All the technologies are energy intensive but financial pangs of the green house technology are not felt as solar energy is freely available in abundant measure, specially in tropical countries. The membrane technology typically requires less energy than the distillation one. Nevertheless, both are energy intensive. The development of techno-economic, large scale distillation processes for sea water desalination preceded those that employ the concept of membrane filtration (reverse osmosis) by about two decades or so. Today, therefore, plants based on either of the two process routes, or their hybrid versions could be found in the world. However, with costs of generating electricity from solar energy coming down, production of potable water from sea or sewage water is becoming a highly competitive process.

Components of Desalination Plants

Any desalination plant consists of the following wings:

- Source of energy to drive the plant.
- Arrangements to receive sea water.
- The plant itself.
- Arrangements to collect the pure water.
- Disposal of solute enriched waste water, called brine, into the sea in an eco-friendly manner. Inherent in the statement is the fact

that only a fraction of the saline water entering the process is converted into pure water and the salt removed is dissolved in the residual sea water, now called brine. Brine disposal is an environmental issue in some areas where the flora and fauna are sensitive to local increase in seawater salinity and increased temperature because the discharged brine is nearly always at a higher temperature than sea water, except in the membrane processes.

The important components of a desalination plant are shown schematically in Figure 1.

Figure 1

Basic scientific concepts in utilization of solar energy for accelerated evaporation

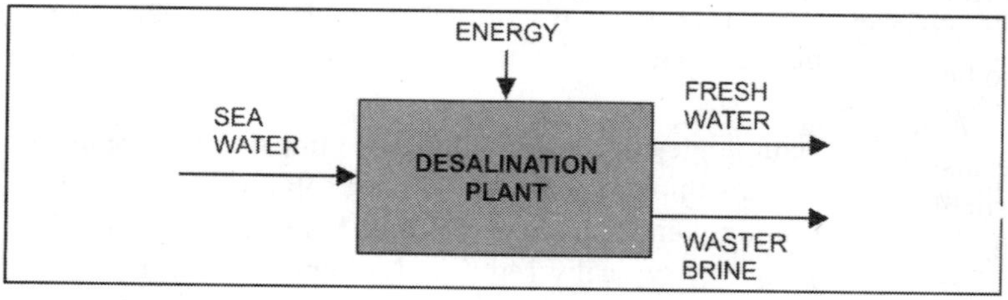

How Does Solar Energy Reach Earth

Almost all the energy that drives the various systems (climate systems, hydrologic systems and ecosystems) originates from the Sun. It is emitted by the Sun and reaches us in the form of electromagnetic waves of sunlight, both in the visible and invisible wave bands. The Sun with a surface temperature of about 6,000°C, (the surface of the Sun is the source of all radiations from it), is located about 14,96,00,000 km away from the Earth. It radiates its thermal energy through a spectrum of waves which travel at a speed of about 2,99,792.5 km/s in all directions in space. The expanding cone of radiations coming in its direction is arrested by the Earth. Since there is absolute vacuum in the space between the Sun and the Earth, there is no loss of energy during its long journey to the Earth. But the energy transmitted, or radiated per unit area keeps decreasing as the rays travel farther away from the Sun, in accordance with the laws of radiation. As the Earth's atmosphere, consisting *almost* wholly of nitrogen and oxygen, is fully transparent to Sun's radiation, the surface of the Earth receives the solar energy in full measure (for those interested in details, it may be stated that minor gases like dioxides of carbon and sulphur, water vapour, and methane present in very minor quantities in the atmosphere do absorb a very minor amount of solar radiation).

What does it mean? Normally, we become aware about the presence of a hot body, close to us, through the radiations the hot body emits. Now conceptualize that three independent columns of pure carbon dioxide, nitrogen and oxygen, each at 1,000°C, can be made to exist. We would not become aware of the presence of hot columns of nitrogen and oxygen as they would not be emitting any thermal radiation but would be able to sense the presence of hot column of carbon dioxide. It is a law of science that only those bodies that absorb thermal radiation can radiate it. If we keep a block of steel or carbon (charcoal), heated to 1,000°C, it would not only radiate heat but would also be visible through its glow.

Among other transparent materials on Earth, glass and plastics are also transparent to it but not water which gets heated up by absorbing the thermal energy. So do the other opaque materials like soil, rocks, metals, ceramics, etc.

How Does the Solar Thermal Energy Travel?

Thermal radiation is one of the fundamental mechanisms of heat transfer. The Sun transfers a part of its thermal energy to Earth through this mechanism. Sunlight, with which we are visibly familiar, is a part of the thermal radiation generated by the hot plasma of the Sun. Indeed, visible sunlight is just a part of the total spectrum of electromagnetic radiation given off by the Sun and composed of waves of different wavelengths. This spectrum is shown in Figure 2. It can be divided into three parts:

- Ultra radiation band of very small wavelengths, (100–280 nm, nano metre). These are not ordinarily visible to human eye. Most of it is absorbed by the upper reaches of the atmosphere.
- Visible radiation in the wavelength band of 380–780 nm. About half of the radiation is in the visible short-wave part of the spectrum. The other half is mostly in the infra-red part.
- Infra-red spectrum with wavelengths between 780–10,00,000 nm (= 1mm). This band is not visible to the human eye. It contains the other half of the energy of solar radiation.

Figure 2

Solar irradiance spectrum above atmosphere and at surface. Extreme UV and X-rays are produced (at left of wavelength range shown) but comprise very small smounts of the Sun's total output. Physicists quantify the solar energy received in terms of watts per square metre.

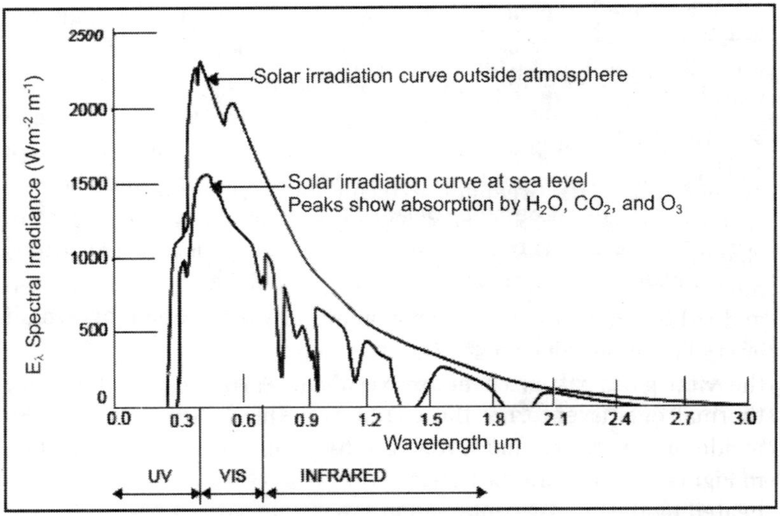

Source: https://wiki.hexagongeospatial.com//images/1/1d/Atmospheric_absorption _sun_illumination_diagram.png

It may be clarified that any material body emits thermal radiation but the wave bands of the spectrum may not be in the visible range. The waveband of radiation emitted by a body is characteristic of its temperature. At lower temperatures, the thermal energy is contained in waves of higher wavelength in the infra-red region. Our own Earth also emits a thermal energy spectrum in the manner of any celestial body. But it is in the infra-red region because the surface of the Earth is at much lower temperature (thermal radiations emitted by a body depend on its surface temperature, irrespective of how high the internal temperature may be). It may also be stated that glass, which is transparent to visible light is opaque to infra-red and ultra violet radiation.

When a beam of sunlight falls on a body, the body absorbs a part of the thermal energy contained in the beam, resulting in the increase in temperature of the body. The thermal energy absorbing capacity of all material bodies is not identical. One can easily experience this effect by trying to walk bare footed on a pitched road on summer afternoon!! But one can walk without discomfort on an exposed platform of wood.

Desalination of Sea Water: Technology Mimics Natural Hydrologic Cycle

Science has tried to mimic the natural process of humidifying air through evaporation from open surface of seas in one region of the world and carrying it to another for its dehumidification through condensation. This process of humidification - dehumidification is sought to be completed within a specially designed enclosure, called a 'Greenhouse'. The Greenhouse technology provides (a) interfacial surface between dry air and saline water, (b) heats up the air in the chamber so that evaporation could occur at fast speeds, (c) pushes the humidified air to another part of the chamber for dehumidification and (d) arranges to collect the condensed water. Since the thermal energy required for humidification is solar, as in the natural process, the vital parts of the chamber are made of glass so that the solar thermal energy entering the chamber with light waves is trapped inside to heat the air and water. The concept of green house is shown in Figure 3. It may be made clear that the inside of the green house is heated by the energy contained in the visible rays as the glass roof is not transparent to the infra-red radiations. Thus, the glass house uses only about half the incoming solar energy to heat it up.

Figure 3

A conceptual model of 'Greenhouse' (Mechell & Lesikar 2010).

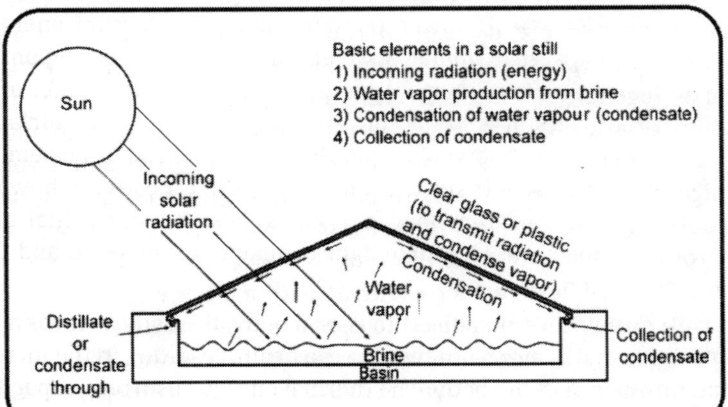

The main elements of a Greenhouse are:

1. The roof of the chamber is sloping and made of glass. The glass house is located in such a manner that its front half faces the Sun throughout the day but the other half is in shade. In the northern hemisphere, the active or solar radiation receiving surface is, therefore, south facing. Glass and transparent plastic sheets are the chosen materials as they let sunlight come in to heat the insides, including water, and then to act as a blanket for the radiation emanating from the heated bodies inside because they are opaque to thermal radiations. The cooling action on the other half of the roof can be enhanced by keeping it covered with hay moist.
2. Arrangements to pump sea water inside the chamber and to let it flow in controlled manner so as to maximize the surface area. It may be done by putting black-coloured stone or metallic spheres on the floor and/or having the water flow in cascade manner inside the greenhouse. The incoming sea water may be separately heated in solar heaters before being pumped in the green house for evaporation. Suitable arrangements are made inside the greenhouse to push the moist air towards the cooler part so as to cause condensation of the evaporated water vapour. In decentralized units, the required electrical energy for pumping water may be generated through photo-voltaic cells.
3. Channels to collect the condensing water and to lead it to storage tanks.
4. Arrangement for suitable discharge of concentrated brine back into the sea, without interfering with the local ecology of the sea.

With this type of technology, about 2–5 litres of water can be harvested per square metre of interfacial area between air and water. There are wide possibilities of using the concept as there is no limit to human ingenuity. One such was developed by Paton in United Kingdom for use in arid, hot desert regions across the world and is presented here as it is gaining recognition in many parts of the world. Its significant feature is that it uses the naturally hot tropical desert air to cool itself down through the process of evaporation of sea water to control the climate in a small encapsulated space. As the moist air moves forward, it is further enriched with moisture produced from sea water and then condensed to produce fresh water. The process evaporates much more sea water than it condenses. The process mimics Nature as evaporation of sea water stores the thermal energy of the Sun in water vapour and releases it elsewhere when moisture condenses to form clouds; the clouds may rain in yet another region and thus bring down the atmospheric temperatures. This system is now known as the 'sea water greenhouse technology'.

Indeed, the same concept is behind the dessert air coolers, widely used in dry and hot summers in northern India to cool the hot air. The concept has independently begun to be used in India to cool down the atmosphere in open air reception venues by creating a very fine spray of water in air.

Sea Water Greenhouse Technology

This technology was specifically developed for those arid and hot regions which are close to the sea with the twin objective of growing vegetables and produce potable water. It is difficult to grow plants in the open fields in hot and sunny regions due to high temperatures and lack of water both in the air and in root support zone of the plants in the ground. Paton in United Kingdom conceived of a unique idea to overcome these difficulties. He adapted the concept of greenhouse to create a temperate climate inside an encapsulated space, using the hot outside air to evaporate sea water from strategically placed surfaces, wetted by sea water, and thereafter by sending the cool and humidified air inside the capsule or the adapted 'Greenhouse'. The environment inside the Greenhouse in terms of temperature and humidity is almost like that in the temperate climatic zones and is conducive to plant growth. It differs from conventional Greenhouses as it uses the hot outside air (and does not use the greenhouse) to heat it. In hot climates, the inside temperature in conventional Greenhouse could become high enough to be unable to support plant life.

The sea water Greenhouse is a method of cultivation that provides cooled and humidified atmosphere within a bounded space and integrates the system to produce potable water. The concept is shown in Figure 4.

Instead of using a Greenhouse to heat the air, the 'sea water Green House technology' directly uses the hot environmental air, brought to temperature by the thermal energy of the Sun naturally. It then recreates the natural hydrologic cycle within a controlled environment to produce cool air and fresh water. The front wall of the set-up faces the prevailing wind (often at temperatures of about 45°C and above) direction in hot desert coasts. In fact, the wall is a honeycomb lattice built of cardboard material with pumped up sea water trickling down the lattice. Fans direct the movement of hot air

Figure 4

Section through sea water greenhouse

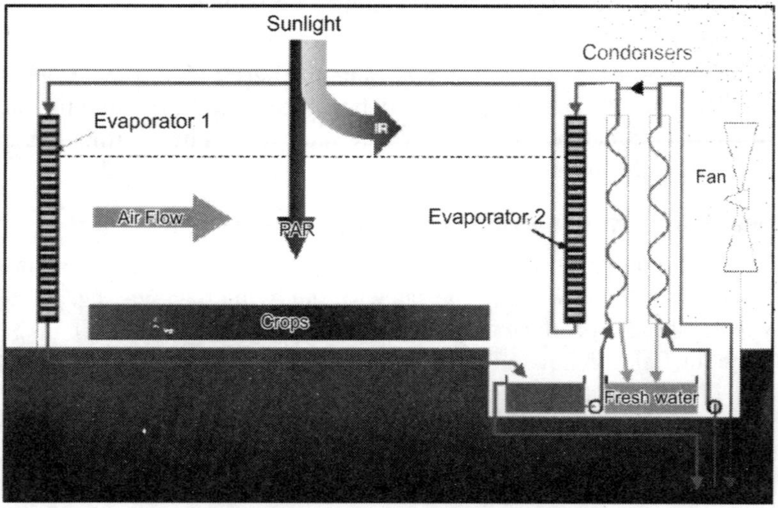

(Paton, 2001: Seawater Green House Development, Oman, MEDRC Project Report As 005b)

through the honeycombed wall. Thus, cooled and humidified, air enters into a large covered space. Sunlight is filtered through a specially constructed roof. The roof traps infra-red heat, while allowing visible light to pass through into the enclosed space. The visible light in the wave spectrum of 400 to 700 nanometres is used by the plants inside the encapsulated space to carry out the process of photosynthesis through which the plants grow. This band of radiation is often called the 'photosynthetically active radiation' (PAR). Sunlight is selectively filtered by the roof elements to remove the infra-red radiation (that does not contribute to photosynthesis but is effective only in raising the environmental temperature). This helps to keep the Greenhouse cool while allowing the crops to grow in high PAR light conditions. The roof has specially designed passes through which the flowing sea water is heated up to pick up large quantities of water as humidity from the sea water. It then flows through a condenser. The condenser is cooled by incoming sea water, drawn from deeps of the sea, where the temperatures are always low. Ordinarily, the surface sea water temperature is about 35°C but 200–500 m below, the temperature drops to about 10°C. The large temperature difference maximizes the amount of fresh water which condenses out of the air stream and is piped to storage. These activities are comprehensively illustrated in Figure 5.

Figure 5

Sea water green house, Paton and Davies (1996)

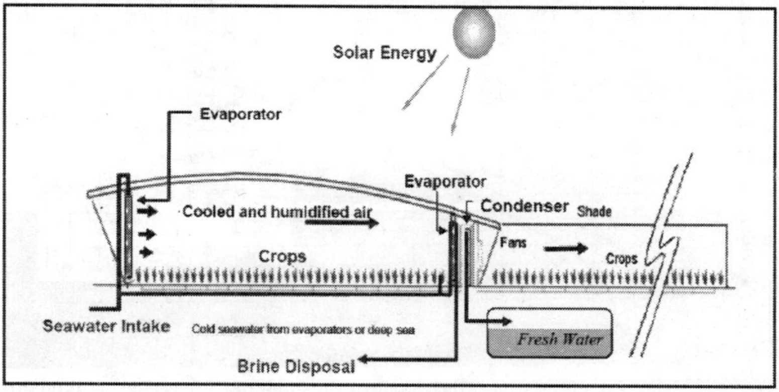

Source: C. Paton, P.A. Davies, The Seawater Greenhouse for Arid Lands'. Proceedings of Mediterranean Conference on Renewable Energy Sources for Water Production. Santorini (1996) 10–12 June.

It is noteworthy that the sea water Green House evaporates much more water than it condenses back into fresh water. This humid air is 'lost' due to high rates of ventilation to keep the crops cool and supplied with CO_2. The higher humidity exhaust air provides some benefit to the cultivation of more hardy crops downwind of the Green House.

The more concentrated salt water may either be further evaporated for the production of salt and other chemicals, or discharged back to the sea with due regard to ecological considerations.

Maximize Condensation of Water from Moist Air

We have pointed out in an earlier chapter that the amount of water required to saturate the air depends both on its temperature and pressure. In other words, air at higher temperature or at lower pressure can hold larger amounts of water vapour. This relationship is shown in Figure 6; atmospheric pressure may be expressed with the help of several units, one bar pressure is approximately equal to the pressure exerted by the atmosphere at sea level. Thus, 100 cu. m of the

Figure 6

Relationship between saturated water density and temperature of the air and the effect of pressure on the amount

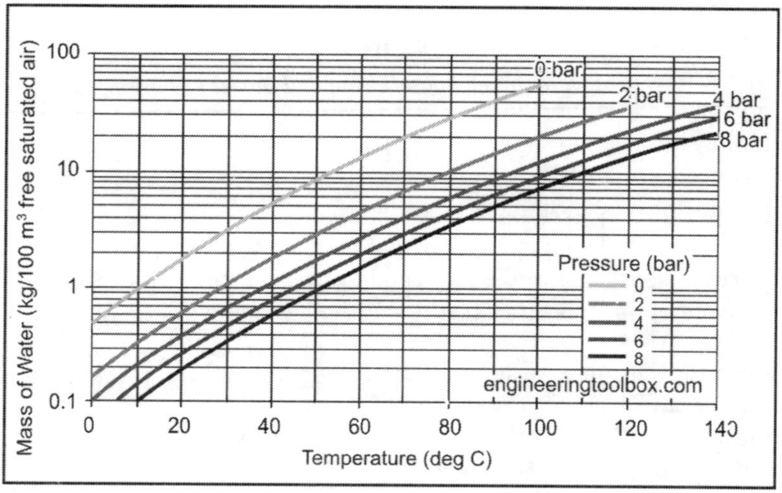

green house saturated air at 50°C contains 18 kg of water. When it is 70 per cent saturated, it would contain about 12.6 kg water. if the Green House is connected to a compressor which compresses the air to about 8 bar pressure and the air exits the compressor at 60°C, the exiting air will be holding 1.6 kg of water but the rest would come out, i.e. about 11 kg of water will come out.

Solar distillation of marine or brackish water, perhaps, holds good potential for exploitation in many parts of India where shortage of potable water is a chronic problem or where the ground water is polluted with ingress of flouride containing water.

Distillation through Condensation of Steam

Despite the technological feasibility of obtaining fresh water through Greenhouses, its important failing is that the process cannot be accelerated. The other scientific process available to produce salt-free water from the sea water is to *cause* steam formation through externally applied heat sources, such as combustion of coal or other petro-fuels as in production of steam for thermal power generation. The production of steam on large scale is an engineering complexity, requiring meticulous designing for high thermal efficiency. Since thermal energy is a cost intensive input, this route of obtaining desalinated water had gained general acceptance in the petroleum rich countries, such as Saudi Arabia and other Gulf countries, where availability of water was the primary issue and energy was available almost just for the asking. Further, thermal energy was also available as waste heat from some industries which was successfully used for the production of steam. Later, the engineered technological version of the scientific knowledge that steam could be formed at considerably lower temperatures when pressure in the steaming chamber is reduced, brought paradigm shifts in the techno-economic feasibility of producing desalinated water, using waste steam at high pressures.

It has been stated earlier that water begins to boil at temperatures lower than 100°C under conditions of vacuum. The relationship between atmospheric pressure and boiling temperature of water is shown in Figure 7. It will be seen from the figure that if the pressure over the steaming surface is sufficiently reduced, steam could form at temperatures lower than 100°C. This phenomenon is being used in desalination of sea water. In this process, sea water is first preheated and is then subjected to a pressure below its vapour pressure causing it to instantly boil or 'flash boil'.

Figure 7

Relationship between boiling point of water and pressure

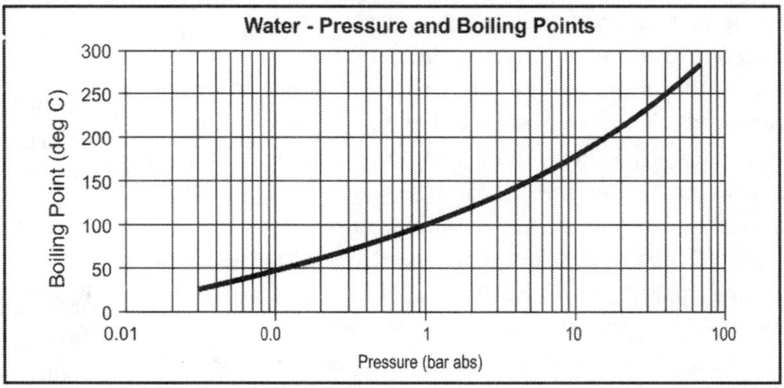

Flash Boiling

In technologically successful processes, sea water is first heated separately in tubes and is then pushed into a chamber at a lower temperature but under a state of vacuum such that the vapour pressure therein is lower than in the heating tubes; as a consequence, the heated incoming sea water instantly evaporates or boils. This manifests in two noticeable events: first, the vapours flash off the heated sea water, leaving the residual sea water enriched in its salt content; and the second, the temperature of the incoming hot sea water falls because the heat of vaporization is extracted from the residual sea water. The important issue is how much of the incoming water at 100°C is converted into steam. In order to quantify the matter, it may be stated that one gram of water requires one calorie of heat to have its temperature raised by 1°C and 550 calories to change into steam. The reverse happens when steam is condensed into water or hot water is cooled down. Therefore in cooling down of the incoming hot water to 60°C, only about 40/550 x 100 = 7.2 per cent of the incoming water would form steam, provided vacuum to the corresponding level is maintained.

The Multi-Stage Flash Evaporation

The process seems inefficient because the evaporation of a small amount causes the temperature to drop dramatically. But the design engineers improved the thermal efficiency of the process by cause flash-distillation to occur in several stages. This is called multi-stage flash, or MSF, evaporation. In this, only a limited amount of steam is allowed to be flash formed in the first distillation chamber under little vacuum. No sooner than the temperature of hot sea water had fallen down to the pre-set temperature, it enters another chamber under greater vacuum so that the process of distillation continues with renewed vigour and so on. Ordinarily there may be five or six chambers – the succeeding chambers being under increasing degrees of vacuum. By the time the input sea water exits from the last chamber under highest vacuum, its temperature has fallen down to about 40–50°C and the residual sea water has turned into brine. This arrangement is schematically shown in Figure 8. The process technology is unique because steam formation and its condensation occurs within the same chamber. Thus, the heat released by the condensation of steam is re-utilized to progressively heat the incoming sea water.

There are two flow paths for the sea water. In the first, it enters the last chamber and successively passes through all the flash boiling chambers; as it does so, it gets progressively heated up from the heat released by the condensation of steam generated by flash

Figure 8

Schematic diagram showing fluid movements in a multi-stage boiler/distillation process

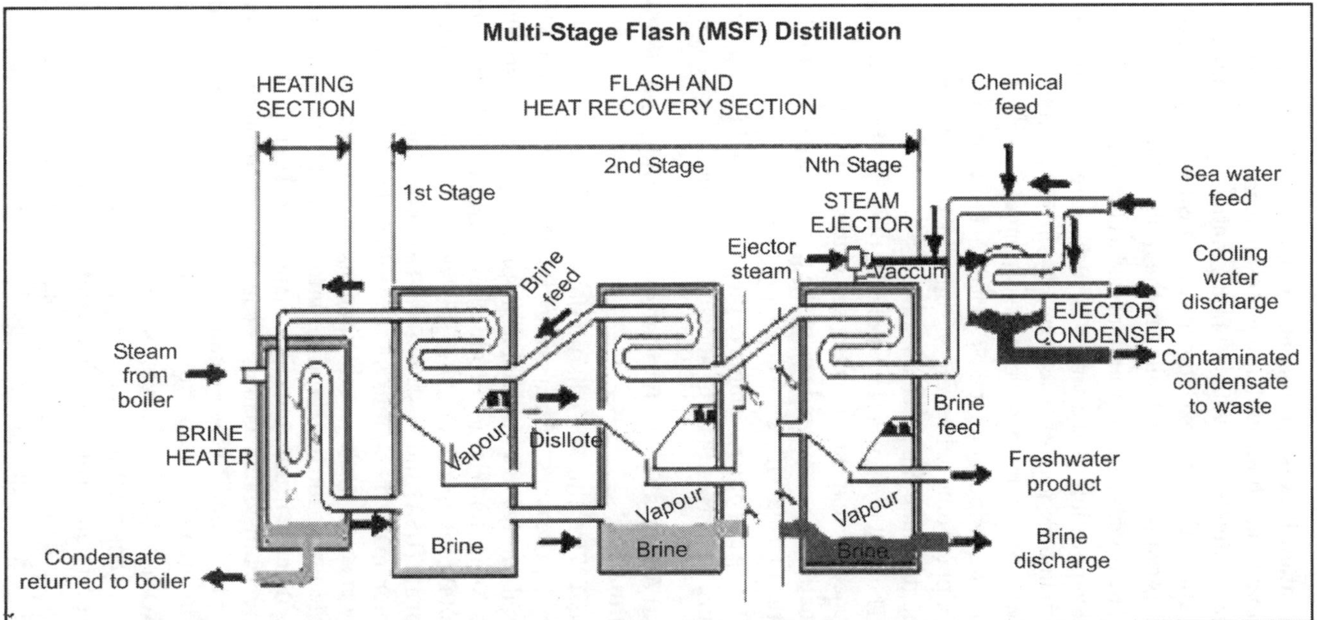

Source: United Nations' Environment Programme.

evaporation. It exits at its highest temperature from the first chamber, only to another independent heat exchanger unit which is heated up by the waste steam, available from another independent process at the plant. In this heat exchanger, the sea water is also pressurized. On exit, it is ready for flash boiling in the first flash distillation chamber.

The process may be summarized as follows:

- Evaporation of sea water is successively done in a number of chambers, each characterized by the level of its vacuum; systems with 5–7 chambers are more common.
- Flash evaporation takes place in each chamber and this can happen only when the vacuum in the succeeding chamber is higher than the preceding one.
- Thermal energy released during the process of condensation is used to heat the incoming sea water or to prepare it for desalination. The repeated use of the thermal energy helps to improve the thermal efficiency of the process.

There are other variants of the process too in which sea water is vapourized instead of being allowed to boil for the formation of steam.

Global Acceptance of Distillation Technology for Sea Water Desalination

In the 1970s, the MSF distillation process was the main stay of thermal processes of sea water desalination in countries where thermal energy was available at practically no cost. Kuwait and Hong Kong had large production facilities for desalination (> 2,00,000 tons per day). Later, the technology found acceptance in California (USA) and Australia where continuing droughts for years together, made it compulsive to produce desalinated potable water at any cost. Australia established six commercial units to produce potable water during the time interval 1997–2007. As the plants were using virgin sources of thermal energy, they remained commercially viable as long as conditions of drought prevailed. But some plants had to face closure as soon as conditions of drought were broken by natural rains.

With recent advances in production of either thermal or electrical energy from solar energy, such desalination plants are once again becoming economically feasible. A plant producing 1,90,000 cu. m/day of fresh water, using solar energy is scheduled to go in production in California towards the close of the year 2016. With the projected commencement of its fifth solar energy-based desalination unit, Israel would be distilling about 5 billion gallons of fresh water annually from the sea using solar energy.

Human Body Fluids Ignite Search for Sea Water Desalination Technologies

Scientists and engineers have marveled at the human body which with the inputs of food and water produces different kind of fluids and excretes them out prominently through sweat and urine, both of which are produced from blood. Physiologists tell us blood carries oxygen, carbon dioxide, nutrients and waste products to liver and kidney to be detoxified and to maintain the water balance in the body by transferring water to and from tissues and produce urine in the kidney. Urine essentially is 96 per cent water – the rest being urea, chlorides of sodium and potassium, creatinine. Its output in a healthy person could be 1–2 litres/day. The other major source of fluid output from a human body is sweat or perspiration which is water based (98–99% water) secretion through the skin. Sweat moves from the sweat glands to the outer surface of the skin through the process of 'osmosis', in which the skin acts as a membrane.

A membrane is a delicate biological sheet that separates one part from another. It may completely block any movement (impermeable) across it or may selectively permit the passage of some substance/s, – now called semi-permeable membrane. Permeable membranes are those that let solvents and solutes pass through them but semi-permeable membranes only let the solvent, water, to pass through.

Sausages are packed in animal membranes and the peritoneum is also a membrane. Scientists began to mimic the functions of kidney in the 1940s. The former was used by the father of dialysis, Koeff of The Netherlands during World War II in carrying out a successful dialysis on an aged lady patient and the latter is being used *in situ* in the latest technology of special kind of dialysis for the benefit of patients with chronic failure of kidney functions.

With seams between different branches of sciences and technology breaking down in the 1950s, rapid strides were made in the production of semi-permeable membranes mimicing the natural semi-permeable membranes of human body. It became a standard medical procedure for patients with kidney malfunction in many hospitals across the world in the early 1970s. This was the harbinger of successful development of large scale sea water desalination technology at the dawn of the 21st century.

Membrane Technology of Filtering Sea Water for Desalination: What is Osmosis

Osmosis is a process which occurs naturally across a semi-permeable membrane which separates two aqueous solutions of different solute concentrations; the solvent from the dilute part selectively moves across the membrane to the other, i.e. the richer one, in order to equalize the concentration between the two compartments. In this respect, the process is closely similar to that of diffusion which also tends to homogenize any concentration gradients. The difference between the two is that in osmosis the solvent molecules move but in diffusion, the solute molecules do. The process of osmosis occurs naturally when plant roots absorb water from the soil. Thus, water from a weaker solution will have a natural tendency to migrate to the solution with higher concentration. This is shown in Figure 9.

Figure 9

Schematically showing the naturally occurring phenomenon of osmosis across a semi-permeable membrane

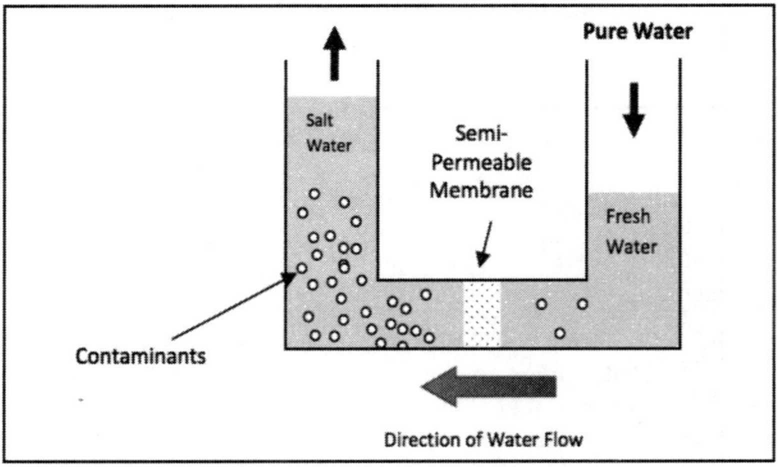

Source: Puretec Industrial Water.

Reverse Osmosis

Reverse osmosis (RO) is the opposite of this. That is, it is a process through which the solvent from the more concentrated compartment is selectively pushed across the semi-permeable membrane to the compartment containing the less concentrated solution. The concept of reverse osmosis is shown in Figure 10. Ordinarily, 5–10 per cent of the solutes contained in the saline water are carried with it across the membrane, i.e. if the solute concentration in feed water is 300 ppm (parts per million), the permeated water could have 15–30 ppm dissolved salt. Since it is not a natural process, external pressure is required to do so.

Figure 10

The concept of Reverse Osmosis

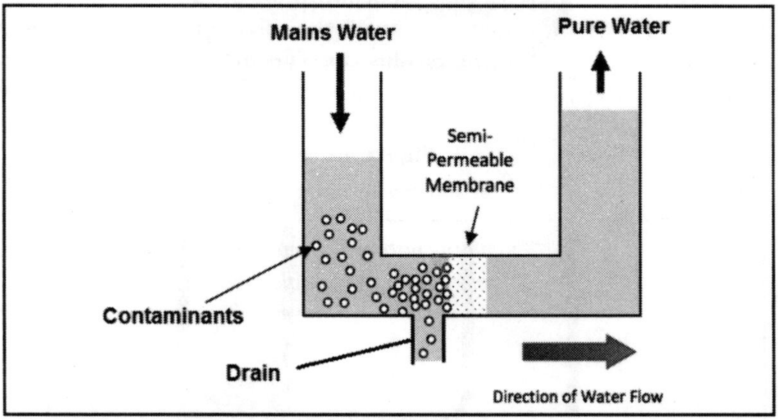

Credit: http://www.hcti.com/sm/aboutro/aboutro.html

In other words, the membrane is permeable to molecules of water but not for molecules of salt. The system thus attempts to reach equilibrium with respect to salt concentration in the two compartments. As water passes through the membrane to the salt solution, the level of liquid in the saltwater compartment rises until enough pressure, caused by the difference in levels between the two compartments, is generated to stop the osmosis. This pressure, equivalent to a force that the osmosis seems to exert in trying to equalize concentrations on both sides of the membrane, is called osmotic pressure. This situation is shown in Figure 11.

When pressure greater than the osmotic pressure is applied on the side of the higher concentration solution, the direction of flow of water through the membrane can be reversed, i.e. water will now move from the higher concentration solution to the lower one. This is shown in Figure 12.

Figure 11

The concept of osmotic pressure

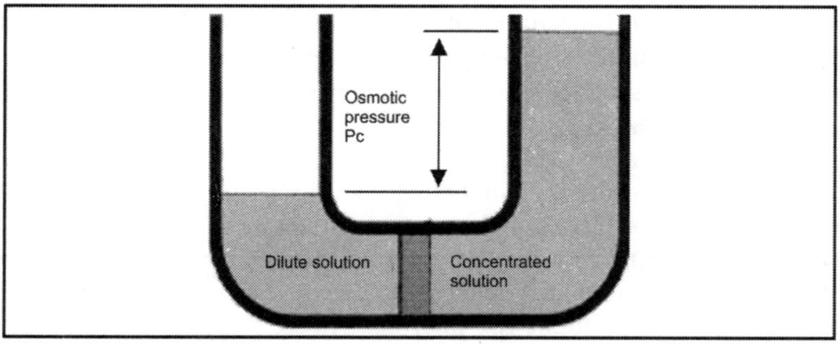

Source: 'Toray' Membrane.

Figure 12

Reverse osmosis

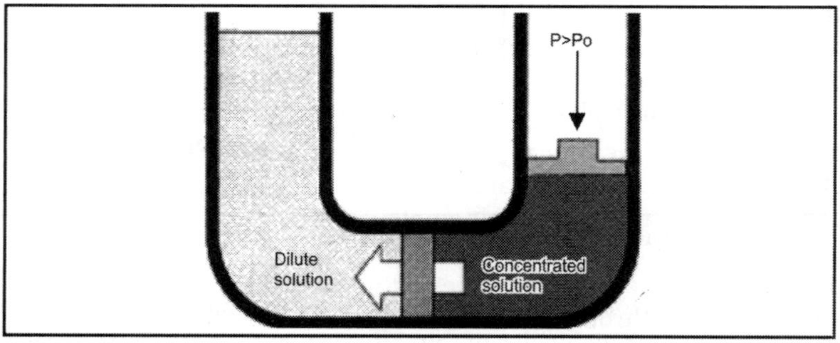

Source: 'Toray' Membrane.

Some sea birds carry out the process of reverse osmosis quite naturally within their bodies. Their physiology is such that one of their body parts desalinates the sea water within their bodies. One of their body parts takes out the excess salt through reverse osmosis and the fluid with excess salt is excreted by a pair of salt glands into the nostrils. Therefore, unlike humans, sea birds have no problem in drinking saline water.

Science was able to mimic this process only after the 1950s. In the beginning, the process required elaborate equipment. Its use for medical purposes for purification of blood (removal of salt, i.e. dialysis) was possible in large hospitals. It was only in the last decade or two that the technology grew to industrial dimensions and was able to desalinate sea water on very large commercial scales.

Reverse Osmosis in Desalination of Sea Water

A wide variety of semi-permeable membranes are being produced. For sea water desalination, the membrane is selectively chosen as would permit passage of water molecules only but act as a barrier for the dissolved salts, organics, oils bacteria including viruses, and pyrogens. The size of each reverse osmosis membrane pore used in the desalination process is about 1/1,00,000th the size of one human hair. The majority of the commercially manufactured reverse osmosis membranes are usually made from cellulose acetate, polysulfonate, and polyamide. The membrane consists of a skin about 0.25 microns and a support layer about 100 microns. The skin is the active barrier and primarily allows water to pass through.

Thus, pure water can be produced from saline water, leaving almost all, > 95 per cent of dissolved salts behind in the so-called 'reject stream' water. The amount of pressure required depends on the salt concentration of the feed water. The more concentrated the feed water, the greater is the pressure required to overcome the osmotic pressure.

Technologically, the process of filtration through reverse osmosis is a marvel of chemical engineering because the system is eliminating solutes which are present on a scale finer than in the colloidal state in the solvent. It employs the concept of what is called 'cross filtration' rather than the familiar and common process of filtration. The solutes are arrested within the filter media. The solution passes through the filter, or crosses the filter and comes out through two outlets: the filtered water through one and the residual, consequently more concentrated, solution through the other. Solute build-up on the membrane is avoided as there is enough turbulence in the water flow to keep the surface clean.

The water remains at ambient temperatures throughout the process. Therefore, on discharge, the reject stream does not cause thermal pollution in the sea but care has to be exercised to ensure that the enhanced concentration is rapidly dissipated so as not to cause any ecological disturbance.

Reverse osmosis process can be adapted for both large and small flow applications. It is suitable for treating brackish underground water, found in many parts of the country, specially in the Thar desert of Rajasthan or water deficient regions of Saurashtra in Gujarat. Brackish water has a lower salt content than sea water. Therefore, potable water can be obtained by applying substantially lower pressures, thereby consuming less energy.

Since its demand of electrical energy has now begun to be met wholly through solar energy, desalination is becoming a standard pre-treatment procedure of natural water for applications in many industries, including thermal power generation and utility boilers, pharmaceutical and beverage industries, etc.

Typical Sea Water Reverse Osmosis Plant

Typical sea water reverse osmosis plant consists of the following units:

- Intake or pumping sea water
- Pretreatment
- High pressure pumps and energy recovery system
- Membrane Assembly for desalination
- Post treatment or remineralization and pH adjustment system
- Storing and delivery system

These are shown in Figure 13.

Pretreatment

Sea water is pumped to a central pretreatment facility, where it is pumped into a distribution system for the purpose of:

- arresting the flow of fine particulate matter (1–5 micron diameter) and biological growth so as to protect the RO membranes from clogging or fouling and high pressure pumps parts from mechanical damage.
- dosing the sea water with oxidizing biocides, such as chlorine to kill bacteria and algae, aeration for the removal of iron and manganese, coagulation for removal of colloids, followed by bisulphite dosing to deactivate the chlorine to protect the membrane from chemical attack.
- Adjusting the pH, hardness and alkalinity of feed water so that the carbonates remain in their soluble carbonic acid form in the reject stream and do not cause any scaling of the system.

High Pressure Pump

Typical pressures used in reverse osmosis plants are:

1. For brackish water range from 225 to 376 psi (15.5 to 26 bar, or 1.6 to 2.6 MPa).
2. For seawater, they range from 800 to 1,180 psi (55 to 81.5 bar or 6 to 8 MPa).

Figure 13

Fresh water from saline water

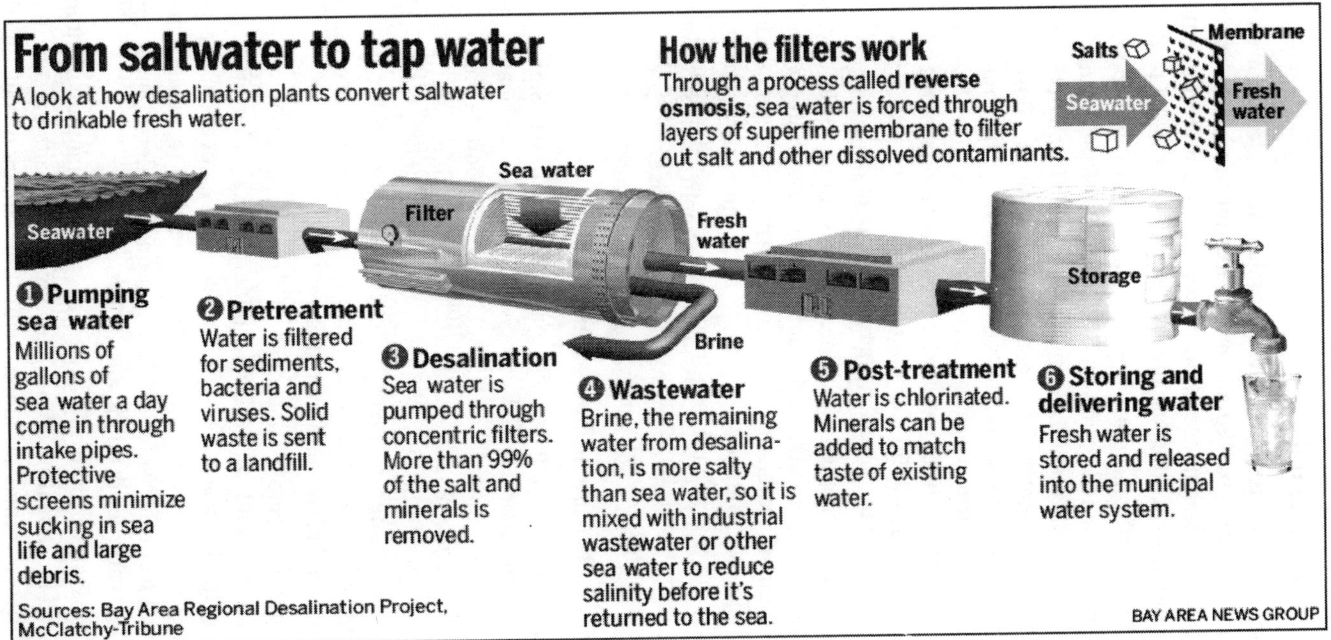

Membrane Assembly

The membranes used in reverse osmosis form a dense barrier layer in the polymer matrix. The membrane assembly consists of a pressure vessel that is designed to allow only water to pass through this dense layer and obstruct the passage of dissolved salts in the feed water.

The membrane must be strong enough to withstand whatever pressure is applied against it. 'Only a part of the saline feed water pumped into the membrane assembly passes through the membrane with the salt removed. The remaining 'concentrate' flow passes along the saline side of the membrane to flush away the concentrated salt solution. The recovery ratio varies with the salinity of the feed water and the system design parameters: typically 20 per cent for small sea water systems, 40–50 per cent for larger sea water systems, and 80–85 per cent for brackish water'. (Wikipedia)

The performance of the membranes tends to deteriorate over time because of fouling and wear and tear associated with dynamic movement of water. In order to increase their service life, plant designers are introducing ultra and microfiltration membrane systems as a standard pre-treatment system. Designers estimate, when introduced, these procedures may enhance the life of the RO filters to over five years.

Remineralization and pH Adjustment

The process does not remove any dissolved gases. Since sea water always contains dissolved carbon dioxide, the permeate water can have a pH value slightly on the acidic side, depending on CO_2 levels in the feed water. Where specifications for potable water demand pH between 6.8 and 8.1, it is adjusted through a liming treatment. Even otherwise, liming treatment is often done to protect the downstream steel and concrete lined surfaces from corrosion. The process is called 'stabilization' of the desalinated water.

Reverse osmosis does not distinguish between molecules of common salt and of desirable minerals, e.g. salts of magnesium, present in the sea water. These also get removed in the process. Remineralization may, therefore, be needed to replace minerals removed from the water by desalination, in order to meet specifications for water for drinking and irrigation.

The water produced from a sea water desalination plant in Israel was used for irrigating crops. The magnesium salts present in the sea water were removed along with the common salt in the RO desalination process. Some crops, particularly those of tomato suffered from deficiency of magnesium which was corrected through the use of suitable fertilizers.

Global Situation with Regard to RO Desalination

There being no alternative to desalination to meet the increasing demands of water, the desalination industry is the future of the world. The increased demands for fresh water through sea water desalination have necessitated installation of units of increasingly larger capacity. It is almost truism to state that as the size of the industrial unit increases, the cost of production comes down. This statement is true for all industries and desalination industry is no exception. Large industrial units are increasingly opting in favour of automation. This is one amongst the factors which influences the cost of production; another is the cost of electrical energy, per se, and the energy consumption per unit of production. These have consistently come down over the years. Moreover, the demand of energy has begun to be met fully through the environment friendly solar energy as it does not add to global warming.

Desalination through reverse osmosis made its entry during the past decade. In view of rapid strides made in the fields of producing RO membranes and in the design of high pressure pump systems the cost of producing desalinated water has dramatically come down. Modern desalination plants have the option to use either of the two technological routes, or their hybrid versions, in order to optimize the cost of production. The cost and economics of the two major competitive production routes for desalinated water remain in a state of flux. They are locale specific and depend on the scale of operation of the chosen route, level of automation, salinity of feed water and the permissible level of salt content in the desalinated water, costs of energy, and of concentrate disposal. (Some examples of salt content in sea water: Atlantic Ocean: 35 g/l, Mediterranean Sea :38 g/l, Arabian Sea: 45 g/l, Dead Sea: 300 g/l). At present, scales are tipping in favour of reverse osmosis (RO) technology as more desalination plants are being built with reverse osmosis (RO) technology.

Research and Development Kindles Future Hopes of Desalination Technology

In view of the great importance of obtaining salt free water through desalination of sea water, research is being aggressively pursued in many parts of the world to develop cheaper membrane materials which are chemically and mechanically sturdy as well. Global research seems to have narrowed down to two new materials – graphene and a molybdenum disulphide (MoS_2) type material.

Graphene is a thin one atom thick sheet in which the carbon atoms are arranged in a honeycomb or hexagonal shaped lattice. Graphene has many extraordinary properties. By weight, it is about 207 times stronger than steel. It naturally repels water, but when narrow pores are made in it, water can permeate through them. This opened up possibilities of using graphene for sea water filtration and desalination, because they are able to let water molecules pass but block the passage of contaminants and substances at much lower pressures than used in conventional RO filters. Graphene filters are set to create new milestones in desalination technologies.

Household Reverse Osmosis Systems

Household drinking water systems are becoming very popular amongst urban Indian populace. The main disadvantage is that they waste a lot of water as they have a low back pressure. The waste water containing the rejected contaminants is nearly always discharged into the house drains. They can discharge the waste water in eco-friendly manner too.

Commercially available domestic RO units carry out a number of operations within the confines of small space. These operations are:

- a sediment filter to trap particles, including rust and calcium carbonate, with or without a second sediment filter with smaller pores.
- an activated carbon filter to trap organics and chloride minerals which may degrade the thin reverse osmosis membranes.
- a reverse osmosis filter, which is a thin film composite membrane, and optionally lamp for sterilizing any microbes that may escape filtering by the reverse osmosis membrane

Latest developments in domestic filters use filters made of nano materials and membranes.

Desalination of Sea Water in India

Gujarat state has the largest sea water desalination facilities in India. Similarly, the Government of Tamil Nadu has been actively encouraging establishment of sea water desalination plants in the state. The state has ambitious plans for establishment of new desalination facilities besides the existing units at Minjur (since 2010) and Nemmeli (since 2013) near Chennai, each of 100 mld capacity. A third RO plant of 200 mld capacity is presently under construction at Pattipulam.

India's first large scale sea water desalination plant was built by Reliance Industries in 1998 at Jamnagar to supply boiler feed water as well as potable water. The capacity saw expansion in 2005/2008 but the new facility now being built up is based on reverse osmosis technology. Together these facilities would be producing 4,00,000 cu. m of desalinated water.

Indian research organizations, Central Salt and Marine Chemicals Research Institute or the Bhabha Atomic Research Corporation have developed technologies of desalination of sea water and brackish ground water and have helped to establish several small plants on a decentralized scale. But the demand for water is so high that establishment of large units is the cry of the day so as to make water available at modest prices. Desalination of sea water has the potential to effectively contribute to India's multi-pronged attack to solve the problem of water scarcity, as environmental concerns for destruction of marine ecology through discharge of brine back in the sea are being scientifically addressed too.

Recycling or Waste Water Treatment

The pinch of the gap between the availability and demand for fresh water has become so acute in the past decades that water managers are looking to reuse the water as much as possible before it returns to the natural water cycle. The technologies of waste water treatment are scientifically proven and allow communities to reuse water for many different purposes, including industrial use, irrigation, and drinking. The water is treated differently depending upon the source and use of the water after treatment.

Since the sources of waste water requiring treatment are varied or the waste waters are differently polluted, there is no single standardized procedure applicable to all types of waste water. Studies have to be carried out for each source to arrive at the best technological solution. Whatever the technological steps of the optimized technology may be, they involve more than one of the following operations:

- Filtration of the incoming used water for the removal of macroscopic solids.

- Primary settling tanks for the removal of readily floatable or sinkable solids; these are skimmed or scraped and removed for separate treatment.
- Coagulation for flocculation and sedimentation for separation of coagulated solids.
- Pre-chlorination for algae control and arresting biological growth.
- Biological Treatment: The waste water is cleaned through a biological treatment method that uses micro-organisms, bacteria which digest the sludge and reduce the nutrient content. Air bubbles keep the organisms suspended and to supply oxygen to the aerobic bacteria so they can metabolize the food, convert it to energy, CO_2, and water, and reproduce more microorganisms. This helps to remove ammonia also through nitrification.
- Aeration for removal of dissolved iron and manganese.
- Filtration through reverse osmosis process for the removal of dissolved chemicals and bacteria, etc.
- Disinfection.

Human society in many countries has accepted to re-use water after appropriate treatment after its beginnings in the United States in the 1930s and its use has since spread to many countries of the world, including India. As of 2010, Israel treats 80 per cent of its sewage (400 billion litres a year), and 100 per cent of the sewage from the Tel Aviv metropolitan area and the treated water is used for irrigation and public works. Israel has developed an innovative method of further treating the 'treated water'. They claim that the quality of reclaimed sewage water can be improved by adding a gravitational filtering step, after the chemical and biological cleansing. This method uses small ponds in which the water seeps through the sand into the aquifer in about 400 days. When pumped out, it is clear purified water.

Australia, Japan and Spain are other countries which are using recycled water in a big way.

Reverse Osmosis Technology in Sewage Treatment

The introduction of special membranes in reverse osmosis technology by the Dow Chemical of the United States in 2015 for the straight production of potable water from sewage heralded the dawn of a new but revolutionary era for waste water treatment. The company has recently built a plant for producing 100 million gallons of drinking water from sewage water. The salient steps in the technology are almost the same as in the sea water desalination technology and are:

- *Screen:* Sewage moves through screens that catch large items such as rags and toys. A 'grit chamber' then uses high pressure air to separate coffee and other detritus.
- *Settle:* The water flows into 'settling basis' where sediments float or sink; mechanical arms ski and scrape the sludge, removing 80 per cent of the solid waste.
- *Spray:* The water is sprayed on to a honeycomb material where bacteria eat the remaining solids. Now it can be pumped into a river or sent to a plant for further treatment.
- *Squeeze:* Filtered solids are heated to 36.6°C and treated for three weeks. Water is squeezed out and solids go to farms to fertilize cotton, alfalfa and animal feed crops.
- *Suck:* The flow enters micro filtration process. It is sucked through thousands of tiny porous straws that remove contaminants.
- *Sift:* Reverse osmosis membranes do the hardest work, pulling out chemicals, bacteria and other super small contaminants.
- *Shine:* The water is finally blasted with ultraviolet light to wipe out microscopic contaminants that might have sneaked through the RO membranes.

Singapore and San Diego have the largest programmes of producing potable water from sewage. Standards of purification of water for potable purposes are laid out by the Indian Bureau of Standards as well as the World Health Organization. All steps of the technology are rigorously monitored, when the end product is intended for human consumption.

The technologies of recycling of waste and sewage waters have come to stay.

References

Bay Area Regional Desalination Project. http://www.regionaldesal.com/

C. Paton, P.A. Davies, The seawater greenhouse for arid lands. Proceedings of Mediterranean Conference on Renewable Energy Sources for Water Production. Santorini (1996) 10-12 June.

Clayton, R. (2011): A Review of Current Knowledge Desalination for Water Supply. Bucks: Foundation for Water Research. URL

Desalination Plants: The Future of California Coast: http://www.redding.com/news/desalination-plants-the-future-of-the-california-coast-ep-457934770-362361021.html

Desalination: International Centre for Water Management> Adapted from Mechell, J.K.; Lesikar, B. (2010) http://www.sswm.info/content/desalination

Mechell, J.K.; Lesikar, B. (2010): Desalination Methods for Producing Drinking Water. Austin: Agrilife Communications. URL

Remote Sensing: https://wiki.hexagongeospatial.com/index.php?title=Remote_Sensing

Reverse Osmosis Basics: http://www.toraywater.com/knowledge/kno_001_01.html

Seawater Greenhouse Development for Oman: Thermodynamic Modelling and Economic Analysis Principal Investigator Charlie Paton Light Works Ltd, UK The Middle East Desalination Research Center Muscat, Sultanate of Oman November 2001. http://www.econologie.info/share/partager/1243550677xX1PKN.pdf

Source book for Alternate Technologies for Fresh Water Augmentation: United Nations Environment Programme: http://www.unep.or.jp/ietc/Publications/TechPublications/TechPub-8d/desalination.asp

What is Reverse Osmosis (PURETEC Industrial Water: http://puretecwater.com/ what-is-reverse-osmosis.html

Index